TELEMATICS APPLICATIONS IN AUTOMATION AND ROBOTICS 2004
(TA 2004)

A Proceedings volume from the 1ˢᵗ IFAC Symposium,
Espoo, Finland, 21 –23 June 2004

Edited by

A. HALME
Department of Automation and Systems Technology,
Helsinki University of Technology, Espoo, Finland

Published for the

INTERNATIONAL FEDERATION OF AUTOMATIC CONTROL

by

ELSEVIER LIMITED

ELSEVIER Ltd
The Boulevard, Langford Lane
Kidlington, Oxford OX5 1GB, UK

Elsevier Internet Homepage
http://www.elsevier.com

Consult the Elsevier Homepage for full catalogue information on all books, journals and electronic products and services.

IFAC Publications Internet Homepage
http://www.elsevier.com/locate/ifac

Consult the IFAC Publications Homepage for full details on the preparation of IFAC meeting papers, published/forthcoming IFAC books, and information about the IFAC Journals and affiliated journals.

Copyright © 2005 IFAC

All Rights Reserved. No part of this publication may be reproduced, stored in a retrieval system or transmitted in any form or by any means: electronic, electrostatic, magnetic tape, mechanical, photocopying, recording or otherwise, without permission in writing from the copyright holders.

First edition 2005

Library of Congress Cataloging in Publication Data

A catalogue record for this book is available from the Library of Congress

British Library Cataloguing in Publication Data

A catalogue record for this book is available from the British Library

ISBN 0-08-044171 8
ISSN 1474-6670

These proceedings were reproduced from manuscripts supplied by the authors, therefore the reproduction is not completely uniform but neither the format nor the language have been changed in the interests of rapid publication. Whilst every effort is made by the publishers to see that no inaccurate or misleading data, opinion or statement appears in this publication, they wish to make it clear that the data and opinions appearing in the articles herein are the sole responsibility of the contributor concerned. Accordingly, the publisher, editors and their respective employers, officers and agents accept no responsibility or liability whatsoever for the onsequences of any such inaccurate or misleading data, opinion or statement.

Transferred to digital print 2008
Printed and bound by CPI Antony Rowe, Eastbourne

To Contact the Publisher

Elsevier welcomes enquiries concerning publishing proposals: books, journal special issues, conference proceedings, etc. All formats and media can be considered. Should you have a publishing proposal you wish to discuss, please contact, without obligation, the publisher responsible for Elsevier's industrial and control engineering publishing programme:

Christopher Greenwell
Publishing Editor
Elsevier Ltd
The Boulevard, Langford Lane   Phone: +44 1865 843230
Kidlington, Oxford   Fax: +44 1865 843920
OX5 1GB, UK   E.mail: c.greenwell@elsevier.com

General enquiries, including placing orders, should be directed to Elsevier's Regional Sales Offices – please access the Elsevier homepage for full contact details (homepage details at the top of this page).

1st IFAC SYMPOSIUM ON TELEMATICS APPLICATIONS IN AUTOMATION AND ROBOTICS 2004

Sponsored by
International Federation of Automatic Control (IFAC)
Technical Committee on Computers and Telematics

Co-sponsored by
IFAC Technical Committees on
- Intelligent Autonomous Vehicles
- Robotics
- Aerospace
- Marine Systems
- Education

Other Co-sponsors
Academy of Finland
Elisa Corporation
Helsinki University of Technology
Nokia
Siemens

Organised by
Finnish Society of Automation

International Programme Committee (IPC)
Halme, A. (FI)   Chair

Digney, B. (CA)
Gray, J. (GB)
Gundersen, B. (US)
Hirzinger, G. (DE)
Kopacek, P. (AT)
Koskinen, K.O. (FI)
Kühlin, W. (DE)
Mitchell, J.R. (US)
Moore, K.L. (US)

Prassler, E. (DE)
Roth, H. (DE)
Salichs, M.A. (ES)
Schenker, P.S. (US)
Schneider, E. (DE)
Schilling, K. (DE)
Spong, M.W. (US)
Wernersson, Å. (SE)

National Organizing Committee (NOC)
Visala, A.   Chair

Hautala, H. Secretary
Kankare, J.
Knuuttila, O.

FOREWORD

The 1st IFAC Symposium on Telematics Applications in Automation and Robotics is the second event in the series started in Weingarten in 2001. There is no doubt that modern telecommunications and information processing technologies provide tremendously increasing new activity possibilities and R&D challenges in the area of remote services, both in traditional industrial automation and non-industrial applications, like mobile machines and robots, traffic systems, constructed infrastructures, buildings and homes. The symposium series has been created to boost these activities within the IFAC community together with other scientific communities active in the field.

Telematics is still a controversial issue in engineering sciences. What it covers or should cover is somewhat fuzzy in peoples' minds. One aim of the symposium is to further survey and takes up those applications of Telematics, which have central meaning in automation and control. In this symposium methodologies, technologies and special application topics related to the area are considered. The topic varies from communication technology oriented studies to various applications fields, like navigation, tele-operation, tele-education, and space and military.

We have also an industrial session where companies have their statements and an interesting possibility to visit an energy company performance centre, where Telematics is in everyday use. Although the program was not massive we hope that the quality and interesting discussions between participants compensated for the high number of presentations.

We hope that the meeting offered the participants interesting scientific ideas and has led to enjoyable memories.

Aarne Halme
IPC Chairman

Arto Visala
NOC Chairman

CONTENTS

TELEOPERATION AND TELE-EXPERIMENTATION OF ROBOTIC SYSTEMS

VIRTUAL LABORATORY EDUCATIONAL SYSTEMS

SPACE AND MILITARY APPLICATIONS

Copyright © IFAC Telematics Applications in Automation
and Robotics, Espoo, Finland, 2004

www.elsevier.com/locate/ifac

TELEROBOTICS;
TOWARDS EXTENDING YOUR SENSING AND HANDS INTO A REMOTE REALITY

Åke Wernersson

Robotics ,Computer Science and Electrical Engineering
SE 97187 Luleå University of Technology, Sweden
ake.wernersson@sm.luth.se, akewe@foi.se

Abstract: The problem addressed is control of robots and/or sensing, in workspaces at remote locations. Topics taken up in this plenary talk includes laser sensing, scene interpretation, telecommands for local autonomy, research issues especially for multirobot systems. Applications includes aerospace testing in north Sweden.

Telecommands are studied for high level control of robots over a communication channel with a non-neglectable time delay, time jitters and variable bandwidth. The sensors onboard the robot are a time-of-flight range measuring laser and a video camera. A few images are used by the operator for interpreting the scene. From the interpretations of the workspace around the robot, the operator specifies interactively the sequence of individual operations during a composite task. Each telecommand is then executed autonomously by closing the feedback loop between the robot and the objects in the surrounding workspace. *Copyright ©
2004 IFAC*

Keywords: telerobotics, telecommands for semiautonomous remote operation, time-of-flight laser, video camera, incremental map building, bandwidth, time delay, time jitter, mutual consistency, graceful degradation.

1. INTRODUCTION – THE CHALLENGE

First, the title may need some explanations; during the autonomous execution of telecommands predictive simulations should be used for critical passages. Some virtual reality software might be used. To stress safety etc. the term "remote reality" is "invented". The plural form hands is used since really useful remote systems are expected to consist of a few tightly coupled robots and/or spatially distributed multisensor systems.

The presentation will take up several topics on control and sensing in a "remote reality". In these few pages we focus on telecommands and the very difficult problem of automatic scene interpretation. More details and other topics are on the web pages [T04, RA04].

The presentation will take up several topics on control and sensing in a "remote reality". In these few pages we focus on telecommands. This is also a way of solving many applications without running into the very difficult problem of automatic scene interpretation. More details and other topics are on the web pages [T04, RA04]. One simple basic fact should be remembered; measured in terms of time delay for human action the size of our Globe is quite reasonable. The round trip for information exchange between Scandinavia and Australia is approximately 0.1 seconds. The concept of Global Personal Robots, GPR's, is a reasonable goal for the near future.

2. TELECOMMANDS - SCENE INTERPRETATION

The principal block diagram for control and sensing using telecommands is illustrated in Figure 1. The reader should observe the association box at the

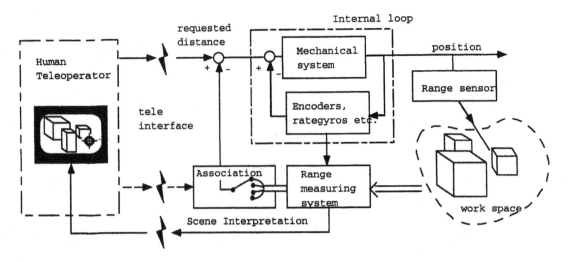

Figure 1 The basic block diagram of telecommands for control of robots and for sensing at remote locations. For different tasks, a mechanical system should be positioned relative to objects in the workspace. The internal loop gives the mechanics a smooth motion. The external geometrical loop is based on non-contact range measuring sensors. For a fully autonomous system the scene is also to be interpreted from the sensor signals i.e. each object is to be associated with the correct signal features. Essentially, using telecommands this loop is closed when the operator selects the association switch.

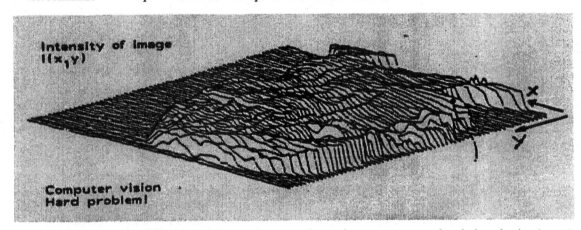

Figure 2 To illustrate the difficulties with scene interpretation a picture was scanned and plotted using intensity profiles I(x,y). For white I = 0 and for black I = 1. Can you interpret the picture? As a clue, what is at the arrow? The picture and more details can be found on the web pages [T04, RA04]. Very few persons have managed with the interpretation. Automatic scene interpretation is a difficult problem and needed for many fully autonomous systems.

lower middle of the figure. In this box a decision is made, automatically or by the operator, on the identity of those objects in the scene that are relevant for solving the task. It is obvious from the block diagram that correct and reliable scene interpretation is essential for autonomous robots.

Figure 2 is a mesh plot of an ordinary photograph. Can you interpret the scene? The difficulty for human interpretation in this case is that the plotting format is, essentially, wrong. For more details, see [T04, RA04]. Summing up some findings, complex problems like automatic interpreting of scenes can be left to the operator in a transparent way. This shifting between different degrees of autonomy yields flexibility.

For comparison, Sheridan [S92] divides the degree of automation of vehicles into four categories: (1) manned vehicles, (2) teleoperators, (3) supervisory controlled telerobots and (4) autonomous robots. In some situations the operator may want to control the robot in detail, in other situations detailed control is unnecessary and only exhausts the operator. One solution to this problem is to use different robot control modes with different degrees of autonomy. The operator chooses control mode depending on the situation. Detailed control is taken care of by the system when possible, and by the operator when necessary. Telecommands falls between level 3 and 4, it is supervisory but with local autonomy during execution.

3. SCENE INTERPRETATION IN FOREST.

Next consider the forest scene in Figure 3. It will illustrate typical properties of laser measurements and modelling for automatic scene interpretation as well as different types of telecommands for control.

The first picture is the intensity of the reflected laser pulse. This is just like an ordinary camera but with a special illumination. The second plot b is range with superimposed equidistance curves.

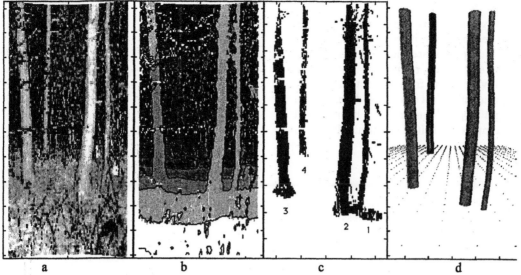

Figure3. Typical laser measurements of a forest scene with several trees and cluttering vegetation. All figures are data projected on a plane. From left to right we have **a**: Measured laser intensity. **b**: Measured range plotted using 12 grey levels. Each grey-level represents 2.5 m in depth and black is 30 m away. As a complement the contours of a 3x3 median filtered range was super positioned on the range image as thin black lines. **c**: The result after the first segmentation step - four trees have been segmented. **d**: Final shape estimation result. The dots indicates the detected ground level at -3.0 m relative to the sensor.

The first part in the processing is a "vertical Radon transform" that gives a number of peaks. The plot in part c is essentially the 3D measurements behind each of the 4 largest peaks. Using robust least square fitting, the resulting geometrical models of the tree trunks are plotted in d.

The forest scene in Figure 3 is next used to illustrate four different semi-autonomous telecommands;
1- The first example is to drive the vehicle in between the trees avoiding obstacles like large stones and inclined slopes. The driver points where he want to pass, or, better the system makes a suggestion and the driver gives an OK.
2- A forest machine with the task of cutting down a selection of the trees. The operator specifies the tree and the machine then executes the task, or better, gives an OK to a sequence of trees suggested by the machine. Using the laser measurements, the machine can make an optimal chopping of the trunk into logs.
3- Mapping for navigation, SLAM, or just for forest inventory.
4- Related is mapping for detecting changes by, say, hidden objects or growing vegetation. Similar but for finding deep wheel tracks in soft soil or detecting damage to the vegetation.

The four semi-autonomous telecommands above were examples illustrating the concepts. However, this rises the question; does not existing autonomous navigation systems already solve these problems? The navigation system in [NDC04] is based on directions to retro-reflective tape and is a mature technology with several thousand industrial users. Making a long story short, there are several major differences between the two cases. Robots to be controlled by telecommands are to operate in highly unstructured workspaces that differs from solving logistic problems in industry. Also, the reflective tapes gives good S/N-ratios that solves association

problems. Several solutions can be transferred between the two cases, but this will be the topic in other publications.

4. SUMMARY AND FUTURE WORK.

Above we outlined telecommands for controlling robots/sensing at a "remote reality". The talk will include several other topics and the reader is referred to the web pages [T04, RA04]. Below we mention some topics for future work.

Telerobotics, without additional tools, is not a good idea for going sightseeing into an *unknown* world. Using only video cameras you may even have difficulties of driving the robot back to the starting position. Some kind of automatic map generation system is needed – compare SLAM. But we should go further using synthetic environments [S04] and computer controlled viewpoints.

For "local operations" some kind of "laser enhanced telepresence" turn out to be very useful. Just structured illumination gives the operator an increased "feeling of depth" in the workplace.

When solving problems in a "remote reality" the advantages of using two or more moving sensor platforms is apparent. This rises the question;
- What can be done with two plattforms that can *not* be done with one at a time?
For future research the research triangle with the interacting parts
- Dynamic models for tightly coupled plattforms.
- Multisensor data fusion.
- Data communication in ad-hoc networks.
is very important.
We needs a *model based framework for telerobotics*. The use is for design of algorithms, for detecting contradictions between sub-systems so we can get at least a graceful degradation, etc.

For the future we also need "hi-tech" applications that serves as technology drivers. One such case is the unmanned European space shuttle Phoenix. Tests of the landing system are reported in [N04]. The vehicle landed within 2 cm from the centreline! Also a network of student project can be the driving force for low budget applications in, say, home based healthcare - compare the MICA wheelchair linked to [RA04].

REFERENCES

[S92] T. Sheridan; Telerobotics, Automation, and Human Supervisory Control, MIT press, 1992.

[H94] G. Hirzinger, et.al.; ROTEX-The First Remotely Controlled Robot in Space, IEEE Int. Conf. Robotics and Automation, San Diego, May 1994

[GS02] K. Goldberg, R. Siegwart; Beyond Webcams; an introduction to online robots, MIT press, 2002.

[T04] http://idefix.ikp.liu.se/rames/telerobotics.html

[RA04] http://www.sm.luth.se/csee/ra/

[NDC04] http://www.ndc.se

[S04] http://www.sne.foi.se

[N04] http://www.neat.se

Copyright © IFAC Telematics Applications in Automation
and Robotics, Espoo, Finland, 2004

www.elsevier.com/locate/ifac

NEW INFORMATION AND COMMUNICATION TECHNOLOGY (ICT) CHALLENGES IN REMOTE AND NETWORKED SERVICES

Jouni Pyötsiä, Dr. Tech

Vice President
ICT Development
Metso Automation
PO Box 310
FIN-00811 HELSINKI
FINLAND

Abstract:
The ultimate goal of remote and networked services is to increase the yield, decrease the total cost of ownership and improve the safety. To meet these requirements we have to closely integrate different kinds of automation technologies, information and communication technologies (ICT) and service business processes.
First, the paper describes Metso's transformation strategy from traditional equipment supplier into a knowledge-based service and performance provider.
Next, the paper presents ICT challenges and solutions from technology and business point of view. These solutions are based on Web Services, upcoming semantic web and smart agent technologies as well as new opportunities in application and business process integration.
Finally, the benefits and needs of these solutions and approach are discussed.
Copyright © 2004 IFAC

Keywords:
Information Technology, Intelligent Machines, Embedded Systems, Maintenance, Life Cycles and Manufacturing Technology

1. INTRODUCTION

Rapid changes and discontinuities in the 21st century business environment will challenge companies. To ensure high flexibility, sustainable growth and profitability companies have to find new innovative business solutions. In many cases technology as such is not anymore sufficient to ensure competitive edge. New innovative business solutions call for strong integration of automation technology, information and communication technology (ICT) and business processes.

Especially the close integration requirements are true in new emerging remote and networked service solutions. Embedded intelligence in different machines and systems gives new possibilities for more automated business process operation over network during the machines and systems life cycle.

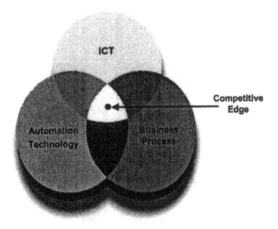

Fig. 1. Ensuring competitive edge in future solutions means more and more close integration of different technologies and business processes.

The new emerging remote service solutions demand that products are transforming into life cycle services and these services are transforming into customers' service processes. Business messages coming from intelligent machines and systems drive these processes utilising embedded intelligence and ICT solutions.

In the future the different collaborative resources like intelligent machines, systems and experts create huge amount of new data and information during the machines and systems life cycles. This information and message flow management and compression to on-line knowledge is also a demanding challenge.

On the other hand, optimisation requirements demand more effective knowledge utilisation and speed up network-based learning during the collaboration between different resources.

To overcome all these demands we have to utilise Web Services, new upcoming semantic web based solutions and intelligent agent approach.

2. METSO'S STRATEGY AND ICT

Metso's strategy is based on an in-depth knowledge of its customers' core processes, the close integration of automation and ICT, and a large installed base of machines and equipment.

Metso's goal is to transform into a long-term partner for customers. Metso will develop solutions and services to improve the efficiency, usability and quality of customers' production processes through their entire life cycles.

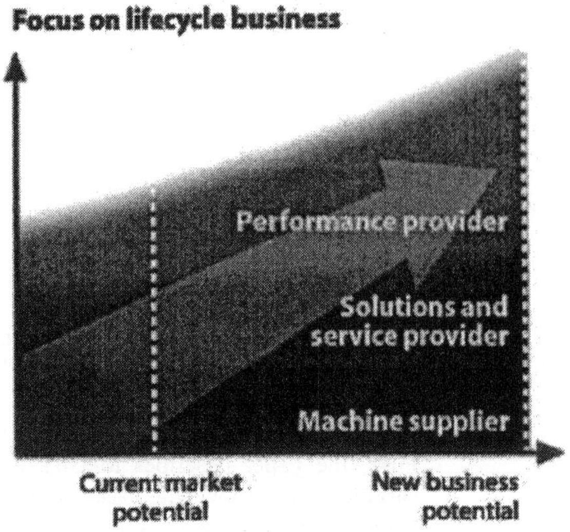

Fig. 2. Metso's large installed base of machines and equipment creates a firm foundation for transformation into after market services.

Close co-operation between the customer makes it possible to optimise entire processes utilising more embedded intelligence. Already in the design phase remote service capabilities can be embedded into machines and processes, which, in turn, form the basis for remote monitoring, process optimisation and optimal maintenance and service solutions. Process optimisation saves energy, raw materials and costs, minimises emissions and environmental impacts and extends process life cycles.

ICT opens up the new possibility for remote and networked service business solutions and presents Metso the opportunity to develop new business models. These models must be based on the life cycle business thinking.

3. ICT CHALLENGES

The problems in today's networked business solutions and remote services start from security. Usually business partners have many point-to-point connection holes into their Intranet used even modem based connections. This usually means low security, difficult management and extra costs for the business partners.

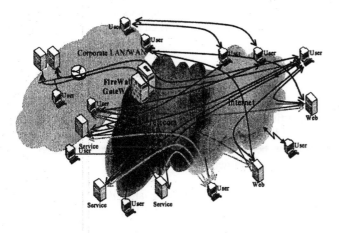

Fig. 3. In many cases today connectivity is a real security risk for different partners in networked business environment.

Another problem is the point-to-point integration between different systems and applications. Business process automation, easy integration and business messages management are very difficult inside the company and almost impossible between business partners.

In today's business environment a key driver for a company's business strategy is its adaptation to a changing business environment. ICT must create a flexible and nimble business architecture based on security and cost efficiency to continuously resolve the highest advantage to the business.

From the technology and business point of view the following ICT challenges have to be met to make the required service business transformation a reality:

- Pervasive Communication:
  - Machine to machine (m2m)
  - Application to application (a2a)
  - System to system (s2s)
  - Collaboration (b2b)
- Network Security
  - High security and confidentiality
  - Easy to build-up and maintain
- Industry Cluster Wide Standards:
  - Security
  - Messages
  - Protocols
  - Collaborative business processes
- On-line Customer Services:
  - Fast response
  - Better focus to real value
  - Network based learning
- Operational Excellence:
  - Strong cost reduction
  - Punctuality in communication
  - Fluent business process operation
  - Transparency

4. EMERGING ICT SOLUTIONS

4.1 Business Hub

Business hub is a solution to provide secure VPN connection between Metso and its customers' intelligent machines and systems. Standard and corporate wide security solution creates fast built, reliable and cost effective solution for customers' integration. Strong authentication, strong encryption and traceability of users and connections guarantee high security.

Enterprise Application Integration (EAI) platform is today's solution for collaboration and business logic (rules and process) modeling in the Business hub. Internal integration is always the starting point for more advantaged collaborated solutions.

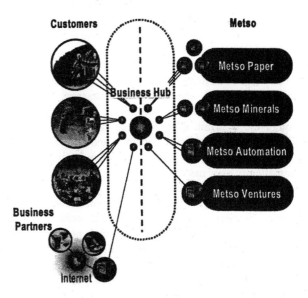

Fig. 4. Business hub based on EAI technology is today's solution for collaboration between customers' intelligent machines and systems.

Main stages in EAI hub roadmap are:

1. Application integration; Stage 1
This level focuses on the exchange of information between databases and business applications via standard interfaces utilizing messaging middleware. The main challenges are well-defined and standardized interfaces, adapters and messages.

2. Process oriented integration; Stage 2
Process oriented integration based on the well-defined business processes and messages. The goal of process oriented integration is to manage business process workflow and message transmission between business applications and sub-processes in networked business environment. Enabling open standards like RosettaNet and ebXML give new possibilities for collaborative process oriented integration.

3. Service oriented integration; Stage 3

Service oriented integration is based on dynamic application integration and common business logic sharing in networked environment. The sharing of large-scale business logic requires a common set of Web Services like UDDI, WSDL and SOAP. Service Oriented Architecture (SOA) has an important role in that.

4.2 Web Services

Web Services are important building blocks for information exchange (SOAP), description (WSDL) and discovery (UDDI) between different resources (like applications, systems, machines and experts) in global network.

Web Services open up new possibilities to move valuable maintenance and process performance information from customer sites to Metso's Remote Service Centers and experts.

The key lies in utilising the intelligence embedded in installed base and automating the message flows between Metso and its customers' applications through the Internet. Web Services based messages and interfaces will allow machines and systems to communicate with each other independently and automatically over network.

Web Services together with EAI workflow and business process tools create powerful vehicle to automate the operation of remote and networked service solutions. Web Services Flow Language (WSFL) and Business Process Execution Language for Web Services (BPEL4WS) give new possibilities to use open XML based standard for business process and workflow descriptions.

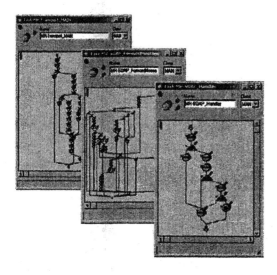

Fig. 5. Ready-made "lego" modules by which it is easy and effective to build up service logic and more automated remote and networked service solutions.

Figure 5 shows workflow by which it is possible to automate different kinds of functions and operations concerning remote and networked services.

4.3 Semantic Web

Semantic web technology led by World Wide Web Consortium (W3C) gives totally new opportunities for building information and knowledge management solutions between different resources in networked business environment.

Integration of Web Services and new enabling semantic web technologies (like RDF, RDF(S), OWL and DALM-S) create comprehensive and more intelligent web services environment.

Resource Description Framework (RDF) provides interoperability and easier discovery between different resources that exchange information on the Web. RDF gives good basis for maintenance and performance information description and classification.

Web Ontology Language (OWL) describes the structure of knowledge and enables knowledge sharing and integration between resources.

DARPA Agent Markup Language for Services (DALM-S) describes the upper level ontology for properties and capabilities and it enables automatically discover, invoke, compose and monitor for web services.

In emerging remote and networked services the information and knowledge discovery and seamless sharing between different networked resources (applications, systems, intelligent machines and experts) are must. To overcome these challenges semantic web approach is an important step to more powerful solutions.

4.4 Field Agent

With the help of agent technology it is possible to develop a simple and advanced performance evaluation and predictive maintenance concept for intelligent machines and devices. This concept is based on smart agents, a network of smart agents and self-learning capabilities.

The agent-based system determines the performance and health of machines and devices with the help of two indices: performance index and maintenance need index. The performance index is a key to evaluating the operation of machines and devices relating to the operational and control performance. The maintenance index is a key to predicting future needs for maintenance.

Field Agent is a software component that automatically follows the performance and health of machines and devices. It is autonomous, it communicates with its environment and other Field Agents, and it is capable of learning new things and delivering new information to other Field Agents. The use of the Field Agent is invisible to the user. It delivers reports and alarms to the user by means of Web Services and new emerging semantic web technologies.

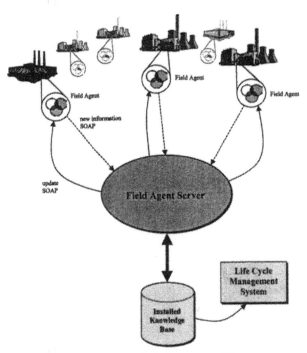

Fig. 6. Field Agent network and server architecture.

The learning of the Field Agents occurs via the Field Agent Server. The Field Agent Server could be a global server that receives new information from Field Agents all over the world, interprets the new information and, if it is useful, sends it to other Field Agents to update their knowledge. The Field Agent Server centralizes information gathering. It also maintains the information on performance and maintenance needs from different plants. This information is valuable for life cycle management and selecting optimal solutions for similar applications.

The emerging semantic web technologies give new possibilities also in implementation of Field Agent concept. Semantic Peer-to-Peer (P2P) architecture provides a direct interaction with other Field Agents over network for learning and service discovery.

5. SUMMARY

It could be said that products and solutions are transforming into services and services are transforming into service business processes in networked business environment.

To make this vision true in remote and networked services we have to closely integrate lots of different kinds of technologies and business processes.

The existing Enterprise Application Integration (EAI) technologies, Web Services and upcoming semantic web technologies give new tools to make this valuable integration a reality.

New solutions based on the previously mentioned technologies will guarantee the increased yield, decreased total cost of ownership and improved safety through more powerful remote and networked service solutions. The key is that the right information is at the right time in the right place in collaborated business environment.

Still a lot of work needs to be done especially in agent-based embedded intelligence and standardisation. More capable intelligence is needed into the machines and systems. To create powerful proactive services we have to get more reliable reasoning and even network based learning to support decision making. On the other hand, standards convergence is a must in a more automated business process operation over network. Otherwise lots of adapters, conversions and transformations shall be made to applications, messages and processes between different business partners.

REFERENCES

Metso Corporation, Sustainability Report 2003, March 10, 2004, URL: http://www.metso.com

Pyötsiä, J., Cederlöf H., Advanced Diagnostics Concept Using Intelligent Field Agents, ISA99, Philadelphia, USA.

Pyötsiä, J., Cederlöf H., Remote Wireless Presence in Field Device Management, ISA EXPO 12-24 August 2000, New Orleans, USA.

Terziyan V., Semantic Web Services for Smart Devices Based on Mobile Agents, In: Forth International ICSC Symposium on Engineering Intelligent Systems (EIS-2004), Island of Madeira, Portugal, February 29 - March 2, 2004

Terziyan V., Kononenko O., Semantic Web Enabled Web Services: State-of-Art and Industrial Challenges In: M. Jeckle and L.-J. Zhang (eds.), Web Services - ICWS-Europe 2003, Lecture Notes in Computer Science, Vol. 2853, Springer-Verlag, 2003, pp. 183-19

Copyright © IFAC Telematics Applications in Automation and Robotics, Espoo, Finland, 2004

ARCHITECTURE AND DATA MODEL FOR MONITORING OF DISTRIBUTED AUTOMATION SYSTEMS

Volodymyr Vasyutynskyy, Klaus Kabitzsch

Faculty of Computer Science
Dresden University of Technology,
D-01062 Dresden, Germany
Fax: ++49 351 463 38460
E-mail: {vv3, kk10}@inf.tu-dresden.de

Abstract: A monitoring architecture and corresponding data model to support fault diagnosis in distributed automation systems is presented. The monitor system consists of a central monitor and a set of the monitoring agents, that cooperate via rule sets and diagnosis results. The data model combines event-based and continuous diagnostic methods and allows to adjust the monitor overhead. The achieved compacting of monitor information on the end nodes along with iterative diagnosis makes teleservice easier and more attractive. *Copyright © 2004 IFAC*

Keywords: remote diagnosis, monitoring, complex events, monitoring agents

1. INTRODUCTION

Modern automation systems like home automation or MES of a factory are typically distributed and consist of heterogeneous components, with increasing role of software part. These properties along with growing complexity lead to fault proneness of such systems during their operation, also if the systems have been carefully tested. As a rule, arising faults are rare, transient and therefore hardly predictable. To cope with such faults, the *monitoring* of such systems is necessary. It is an important part of diagnosis, that helps to indicate possible faults and to avoid more critical fault effects in the future.

Known event-based monitoring systems like ZM4/Simple (Dauphin, *et al.*, 1992), GEM (Mansouri-Samani, 1995), HiFi (Al-Shaer, *et al.*, 1992) etc., are oriented on special system architectures and require a lot of a priori information about investigated systems. This leads to high costs for creating of the telematics knowledge base during installation of the monitoring system or its adjusting by system reconfiguration. In addition to it, exact information about the behavior of some components is often not available during the system design. This information must be obtained during system

integration and operation for tuning of diagnosis system. Therefore monitor architecture must support additionally iterative learning, automated or with human observer, and the mechanisms for easy reconfiguration.

Automation systems combine discrete and continuous processes. Thus the event based monitors mentioned above may also profit from highly developed diagnosis methods for continuous systems, see (Simani, *et al.*, 2003), like diagnosis based on linear and nonlinear models of systems, fuzzy and neural algorithms etc. Although these methods are not considered in this paper in details, the possibility of their coupling with event based methods will be shown.

The iterative diagnosis requires efficient *interaction* with human experts and powerful learning algorithms. The next important point is an optimization of monitoring *overhead* in sources of log data, communication media and data users. Two variants of the distribution of monitoring overhead can be distinguished from these two points of view. In the first case, all log data is collected on the central monitor and analyzed there, in most cases offline. This may provide the full flexibility of data manipulation, but leads to overload of communication medium, that is an expensive and critical resource. The intensive stream of monitoring data can seriously influence total system performance, causing new faults as result of network overload (Kotte, et al., 2002). The contrary approach uses distributed intelligent agents (Köppen-Seliger, et al., (2003), Munz (2001)), that provide all diagnosis activities directly at the end nodes. This is possible due to the trend to more intelligent end nodes of automation systems, allowing to place more powerful processing algorithms. Distributed data processing reduces the network load, but may bring poorly predictable diagnosis results and end node monitoring overhead. Also control possibilities of human experts and using of their intuition are then restricted.

This paper considers the middle way, intended to solve the problems mentioned in the paragraph above. It is asserted, that the preprocessing of monitoring data may be effective due to their high redundancy, conditioned by monotony and periodicity of underlying processes. Simultaneously, the human expert must have the control over diagnosis activities. That's why the monitoring architecture is proposed in this article, that allows to optimize the monitoring overhead in distributed systems and adapt it to diagnosis needs. It uses the monitoring agents, adjustable thru the sets of rules sent from central monitor. The architecture uses a special extendable data model combining detection of complex events with continuous signal diagnosis methods. In that way the process of monitoring and diagnosis are conjugated. The monitoring system is implemented completely in Java and has been tested on some real applications and simulations at the Dresden University of Technology.

The rest of the paper is organized as follows. Section 2 describes the proposed monitoring architecture. The description of corresponding data model follows in section 3, demonstrated on the examples from home automation. Finally, the described approach is evaluated, the perspectives of applying and future work are stated.

2. ARCHITECTURE OF DISTRIBUTED MONITORING SYSTEM

The monitoring system consists of one central monitor and several monitoring agents as shown in fig. 1. Monitor and agents communicate via local network or internet. Monitoring agents access log records from investigated system over interface to monitor sensors, that are system specific sources of monitoring records. Sensors can be instrumented in different ways: as event notifiers, checkpoints integrated in application program code, different buffers, databases etc. Monitoring agent can also access results from the underlying monitor agents, so that hierarchical monitoring systems can be built.

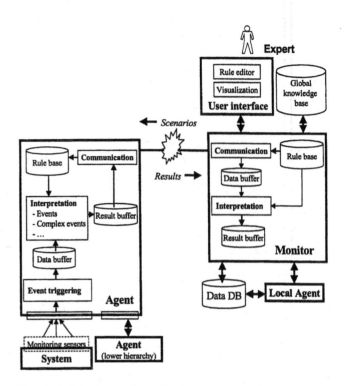

Fig. 1. Architecture of monitoring system.

The monitor sends requests to agents in form of scenarios. Scenarios are a formal description of log data structure, diagnostic rules and diagnostic steps. They also include the communication modes, that describe the cooperation between agent and monitor, i.e. when and which resulting data should be sent from agent to monitor. There are the following modes available:

- in definite time periods, for example each hour;
- by request from monitor;
- by occurrence of definite event. For example, some critical events must be sent to monitor immediately;
- by overflow of data or result buffer.

So, the communication modes are strictly limited and therefore the communication overhead can be better controlled.

Scenarios contain also the status of rules, that defines, whether the results of rule firing should be sent to the monitor. In most cases, it suffices to send only the most relevant high order events. Data model for description of diagnostic rules and results is described in the next section in details.

When the monitoring agent have received the scenario and saved it in its rule base, it starts processing of log information coming from monitoring sensors. Log records are triggered and interpreted according to the scenario rules. Interpreted data come to result buffer and are then sent to monitor according to communication mode and rule status. The obtained results, in their turn, are interpreted on the monitor side. More complex rules and learning methods can be used here, because the data stream there is much more compact and more resources are available. Diagnosis results are passed then to the user interface, that presents them to human expert. Depending on diagnosis results the expert can modify scenarios and send them again to agents and so on. In this way the iterative diagnosis process is supported. Monitor can also use data from the local agents, for example, in case of offline diagnosis.

3.   DATA MODEL

Presented data model describes the structure of log data, diagnostic rules and their results in one uniform system. This helps to evaluate and control the monitoring overhead. The model extends the notations of complex events like GEM (Mansouri-Samani, 1995) with process diagnosis functions. It consists of elements, combined in hierarchies, and is constructed according to the principles of object-oriented programming. The simplified class diagram of the model is shown on fig. 2. A formal language used here is adapted for better comprehension of the reader. In implemented tool the rules are put in over more native visual interface. The elements of data model are described below.

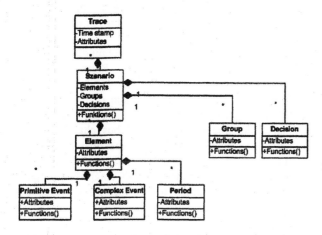

Fig. 2. Basic elements of data model.

Trace presents a log source as a chronologically ordered stream of log records. Each record in a trace possesses several attributes, for example the program module identifier or variable value. In this way the trace becomes independent from source or specific storage form like a text file or a database. The trace presents a $(N + 1)$ – dimensional state space, where N is the number of attributes. One further dimension is time, which is as a rule discrete and not equidistant.

The rules for trace analysis are combined in **scenarios**. As mentioned in section 2, scenarios combine the elements used in the same cases. A scenario represents therefore one diagnosis hypothesis. Several, among others also competing scenarios may be produced during diagnosis. The scenarios are stored in XML-files, that allows the easy data exchange with other development tools. For instance, the scenarios can be automatically generated from UML diagrams (Matzke et al, 2003) or from development information databases like LNS-databases (LON).

The original monitoring events are abstracted in higher order elements like events and time periods.

Primitive events are trace records, that are relevant for a diagnosis hypothesis and have in that way a proper semantic meaning. The primitive events are defined through the attributes of trace, for instance:

Req_17:= (Trace.Source=17 AND
Trace.Msg='Request')

This rule describes the events from the node with ID = 17, that contain the message "Request". Primitive events are used for quick triggering and filtering of log data and are the basis for more complex events.

Complex events combine several events (among others also other complex events), that stay in proper relations to each other. They personify in this way the causal relationships in a system and unite the events in more complex structures, so that analyzed data amount is reduced. Let A and B be primitive or complex events, then following relations can be used in the definition of complex events, compare Mansouri-Samani (1995):

- A; B – event A must precede event B (sequence);
- A | B – one of two events may occur (branching);
- A ~ B – events A and B may occur in any order (parallel execution);
- [n:m] A – event A must appear from n to m times (iteration);
- ! A – event A may not occur (exclusion).

For example, the following expression describes a complex event, that represents a successful transaction, consisting of two activities: sending of request Req_17 and receiving of response Resp_17. The transaction timeout equals 30 seconds:

TA_17:= (Req_17; Resp_17).(Duration < 00:00:30)

The description of causality by complex events is equivalent to such descriptions as timed automata, Petri nets or causal trees, compare fig. 3. The model proposes at the same time more compact, powerful and simply extending presentation. For instance, further event attributes can be simply introduced, that should need additional extensions in Petri nets.

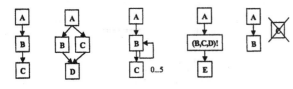

A;B;C   A;(B|C);D   A;[0:5]B;C   A;(B~C~D);E   (A;B)!C

Fig. 3. Relations in complex events in form of automata.

The violations of event rules, also called **outliers**, may indicate possible faults or their causes. That's why they may play an important role and are especially processed. For example, when in the last rule TA_17 the last event Resp_17 does not occur, an outlier is thrown, that represents an incomplete transaction. Depending on protocol details, such outlier can be interpreted as a fault ("disrupted transaction") or as a symptom for network overload, since several such disrupted transactions can be tolerated by the automation system.

Time periods are generally defined as periods of the trace, that possess definite properties. They may embody some temporal properties of the system like:
- System configuration. For example, the time period from 9:00 till 18:00 from Monday till Friday represents the time of human activity in the office, described by such rule: Per_WorkDay:= (9:00-18:00, Mo.-Fr.).
- Overall system state like the period with high network load: Per_High_Load:= (TA_17.Duration>10) AND (TA_17.Frequency > 20)
- Control loop state as the period of transaction, i. e. between beginning and end of a transaction: Per_TA:= (TA_17).Period

Such time periods allow to represent all processes of different nature at the unified time axis, so that the causal relationships may become obvious. The rules can be simpler as in poor event description models.

All elements possess a set of primary and secondary attributes. **Primary** attributes are contained in the source data and represent explicit knowledge of the expert about system behavior. These are for instance the log source or the time stamp of event occurrence.

Secondary attributes, or functions, are omitted on the basis of primary attributes. They can be changed during diagnosis, hence they depend on diagnosis hypothesis. An example of the secondary attribute is a class that describes network load in proper time period ("high load" or "low load"). This class is calculated on the basis of transaction frequency in network channel like that:

Class_Load: = { "high" : TA.Frequency $\geq$ 30;
$\quad\quad\quad\quad\quad$ "low" : TA.Frequency < 30}

Secondary attributes produce additional projection in trace data, so that a deeper view of relationships is

created. They are calculated dynamically and can be added online during diagnosis if it is necessary. Once capsulated in attributes, different model based and model identification methods may be used for diagnosis, such as statistical clustering, neural nets, fuzzy classifiers etc. To be placed in the remote agent, the routines implementing these methods must fulfill the requirements on monitoring overhead, that can be checked by the sample run of the routine with using of typical historical data. The heterogeneous values may be compared with each other by classes obtained from attributes.

Complex events, time periods and functions are internally organized as tree structures, that allows the quick and online capable search. But this requires also the corresponding tree structure of rules. These restrictions are checked in the rule editor.

Further elements, namely groups and decisions, are introduced for purposes of better structuring and comprehension for expert.

Groups are parts of the investigated system or sets of attributes that possess some common properties. They restrict and subdivide the validity space of the rules to make the search for relationships easier and more automated. As an example the group of transactions is introduced, that is produced by different heating controllers in rooms of one house:

Gr_TA_17 : = Group (TA_17, GroupingParameter = Room)

This can be used for comparison of rooms of one house. Further examples of groups are all devices, that communicate on one fieldbus channel, or all devices of an automated house, that possess the same functionality. So, the diagnostician must just define the group, that is relevant for his diagnosis purposes. The further search inside of group proceeds automatically.

Decisions complete data model with expert rules in the form of If...Then statements. For instance, the following rule:

IF TA.Duration.Max>00:00:20 THEN
$\quad\quad\quad\quad$ Residuum:='Problems in communication'

indicates the communication problems and fires when the transaction duration in some transactions exceed certain limit. Here is an example of more complex decision:

IF AllNetworkEvents.Frequency>30 AND
$\quad\quad\quad\quad\quad\quad$ PID.Frequency>20
$\quad\quad\quad$ THEN Residuum:='Oscillations in PID-controller caused by network overload'.

The rule indicates, that if the oscillations in PID control loop are accompanied with the high frequency of messages in the network, then these oscillations may be caused by the message delays.

Ideally, there can be constructed a complex decision, that indicates the normal system behavior ("Everything is OK"). In that case the monitoring overhead on communication medium is minimal, but not the overhead on agent side.

Returned answers from monitoring agent repeat structure of corresponding rules with the difference that, as a rule, only high order elements are transferred. The results can be then presented to human expert online or offline, for example as a Gantt diagram in the diagnosis tool "eXtrakt" (Kotte, et al., 2002), as shown in fig. 4.

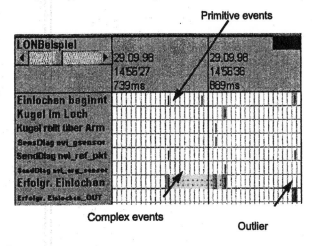

Fig. 4. Presentation of primitive and complex events as Gantt diagram.

4. USAGE OF DATA MODEL

With help of the data model, the queries on monitoring data can be produced. The diagnosis proceeds iterative, starting with simple events describing simple transactions. These events can be generated automatically on the basis of protocol details. Then the queries on different subsystems can be produced, depending on the purposes of diagnosis. For example, the behaviour of a control loop, the behaviour of the household appliances in a room or the behaviour of the whole heating system of the house can be investigated. The results of a diagnosis step can be used later in further diagnosis steps. The queries can be entered in the comfortable way directly in GUI of development tools.

As we have stated in Introduction, application of proposed monitoring system promises more compact and adjustable transfer of monitoring data. The compression rate is larger when only elements of higher order are transferred. This rate depends in general on the purposes of the diagnostician, properties of underlying processes like periodicity, ratio of explicit, available explicit and implicit knowledge about system, fault frequency. It is clear, that the compression rate may grow during the diagnosis, when more explicit information about system becomes known. The compression rate of 5 up to 20 was achieved in tested real applications with

admissible monitoring overhead on the agent side. The choice of proper ratio between monitoring overhead at communication medium and on agent side is not a trivial task. Obtained experience shows that this choice should be made iterative, starting with few rules describing most frequently appearing cases. Usually, only a few iterations are necessary to achieve desirable monitoring overhead. The diagnosis and monitor adjustment can proceed offline as well as online ("on-the-fly") using the same data model.

Described data model is general enough to cope with heterogeneous nodes in the automation systems. It unifies the diagnosis and is easy extendable on the further diagnosis methods. Thanks to the common trace interface, rules do not depend on the source data format and can be reused in different application domains, if the definitions of primitive events have been adapted to format of monitoring records. The model can use the á priori information as well as newly obtained results of monitoring and human experience. An example of using design information is shown in (Matzke, et al., 2003), where UML – diagrams are used to produce diagnostic rules.

The monitor and agents are implemented in Java, so the monitor system may be used in different applications using this language. The OSGI – initiative is interesting in this connection, because it proposes universal interfaces for using Java with different automation systems. On the basis of data model, the specific monitoring agents can be produced in native programming language of the system and placed directly in the automation nodes. This would bring further performance gain, retaining the unified data structure and communication interfaces.

5. CONCLUSION AND FUTURE WORK

The architecture for monitoring of distributed heterogeneous automation systems together with the corresponding data model are presented. They combine description generality with adjustment of monitoring overhead. The data model allows to use the different diagnosis methods in one diagnosis system. The compacting of monitor information on the end nodes along with iterative diagnosis would make teleservice easier and more attractive.

Future work will concern learning algorithms on the monitor and agent sides and the efficient distribution of the learning activities between these two parts. Another task is the representation of diagnostic results to the human diagnostician.

REFERENCES

Al-Shaer, E., H. Abdel-Wahab and K. Maly (1999). HiFi: A New Monitoring Architecture for Distributed Systems Management. In: *19th IEEE*

International Conference on Distributed Computing Systems. May 31 - June 04, 1999 Austin, Texas.

Dauphin, P., R. Hofmann, R. Klar *et al.* (1992). ZM4/SIMPLE: a General Approach to Performance-Measurement and -Evaluation of Distributed Systems. In: *Readings in Distributed Computing Systems (T. Casavant and M. Singhal, eds.),* IEEE Computer Society Press, Los Alamitos, California, 1992, Chapter 6, pp. 286-309.

Köppen-Seliger B., S. X. Ding and P. M. Frank (2002). MAGIC - IFATIS: EC-Research Projects. *In: Proceedings of 15th IFAC World Congress, Barcelona, 2002.*

Kotte G., K. Kabitzsch and V. Vasyutynskyy (2002). Diagnosis in MES of Semiconductor Manufacturing. In: *Advanced Computer Systems, 9th International Conference, ACS'2002,* Miedzyzdroje, Poland, October 23-25, 2002, Proceedings Part 1, pp. 223-238.

LON: http:\\\\www.echelon.com

Mansouri-Samani, M. (1995). Monitoring of distributed systems. PhD Thesis, University of London.

Matzke F., V. Vasyutynskyy and K. Kabitzsch (2003). UML Specification Based Fault Diagnosis on Embedded Systems. In: *Proceedings of Fourth International Conference on Industrial Automation, Montreal, Canada, 9-11 June 2003.*

Munz, H. (2001). The State of PC Based Control. In: *Proceedings of 1st IFAC Conference on Telematics Applications in Automation and Robotics, Weingarten, Germany, pp. 179*

OSGI: http://www.osgi.org/

Simani, S., C. Fantuzzi and R. J. Patton (2003). *Model-Based Fault Diagnosis in Dynamic Systems Using Identification Techniques.* Springer Verlag.

Copyright © IFAC Telematics Applications in Automation
and Robotics, Espoo, Finland, 2004

ELSEVIER
IFAC
PUBLICATIONS
www.elsevier.com/locate/ifac

A CONSULTING MODULE IN ROOM AUTOMATION

Alexander Dementjev, Klaus Kabitzsch

Dresden University of Technology
Institute for Applied Computer Science
D-01062 Dresden, Germany
Fax: ++49 351 463 38460
E-mail: {ad14, kk10}@inf.tu-dresden.de

Abstract: In this article the intermediate results of the work within the research project „Knowledge-based Services in Building Management" (WiDiG) are described and the concept of a consulting module in room automation is presented. The tendencies of further development of algorithms for the room automation are considered as well. The main goal of the joint research project „WiDiG" is to combine the present room automation with facility management systems, establishing one integrative concept for building rationing. *Copyright © 2004 IFAC*

Keywords: building automation, knowledge-based systems, fuzzy systems, interconnection networks, logical control algorithms

1. INTRODUCTION

The building management faces more and more new challenges today. Demands on comfort and flexibility of the building exploitation along with the efficiency and economy of energy become more and more important. The building automation systems are composed of networked components for controlling and automation of functions inside a building. Therefore more complicated tasks have to be solved by designers, system integrators and maintaining companies to meet these requirements.

On the other hand it is possible to shift the intelligence into the field devices in each room due to fast development of field-bus technology. These new technologies lead to new automation approaches and to new solutions of raised tasks (see (Kabitzsch, *et al.*, 2002a)).

The solutions based on data base technologies gain more importance in the building automation. The requirements of facility management at all levels of the building automation have been not sufficiently considered today (Schach, *et al.*, 2001).

The present state of the art in the building automation has following features (Kabitzsch, *et al.*, 2002a):

- Using and further development of the field-bus technologies including integrated circuits, transmission techniques and micro controllers;
- Integration of micro controllers in communication systems;
- Automation functions (evaluation, control, regulation, optimisation, adaptation) are decentralised and distributed between sensors, actuators and controllers;
- Sensors must be no more multiply installed for each fulfilled function (they can be used commonly by various service systems and applications as well);
- The quantity of the acquired data is significantly increased;
- Thereby new room services can be implemented.

So new solutions appear helping to rationalize the maintenance for long periods of time with one's limited resources. Thus building operation costs can be decreased, operating can be simplified providing more comfort. Investigations in this direction are being held nowadays at many research institutes. In (Laukner and Knabe, 2001) an intelligent single-room control is proposed, controlling the air-conditioning and room temperature depending on room occupancy. It also provides the air exchange required by the hygienic, healthcare and physical reasons. In (Kuntze and Nirschl, 1998) an intelligent controlling component for optimisation of the heating and air-conditioning controlling values in dependency on the actual climate and room occupancy situation is described.

In (Tamarit and Russ, 2002) the project "Smart Kitchen" is presented that has the aim to create an application integrating system. That system should be able to recognise situations by means of a large number of sensors, to estimate them and to react via corresponding actuators. Such systems should be able to fulfil the following basic functions: perception and identification of the actual situation in a room and preventive reaction according to the recognised situation.

Remote access and remote control in the room automation gain more importance today (see (Tarrini, *et al.*, 2002)). In (Corcoran, *et al.*, 1998) possible variants of the access to various house systems are described: web-browser with desktop PC, Java-phone terminal, TV-set with remote control or PDA. A PC or PDA with Internet access serves usually as a control interface for the user.

The present state of the art does not offer sufficient possibilities for the user to be included in contemporary resource and energy management in the building automation. There is a lack of prediction algorithms illustrating the customer various activity alternatives and their consequences. The system uses the subjective demand of costumer on comfort, and returns the corresponding costs as output. In that way the consumer may find his own individual compromise between comfort and costs.

One possibilities to achieve this goal is integrative room management system made by one equipment manufacturer trying to optimise controlling of existing devices. This is certainly a very proprietary solution. Another alternative is open architecture, that allows to implement an additional optimisation module communicating with all the controllers, sensors and actuators within one room. This way is followed by developing of the consulting module at the Institute of Applied Computer Science (Dresden University of Technology) within the WiDiG project on the basis of the LONWORKS® Technology and open OSGi specifications and products. The difference to existing solutions supposed to be system integrating but proprietary implemented, consists of:

1. More intensive communication with the user by means of a system integrating open platform. Consulting module is not simply an automatic system – it fulfils consulting functions as well. For instance, it can be the identification of incorrect controlling situation in sun-blinds and warning of users or recommending them to set the correct controlling value.
2. Open and convenient service by means of Open Service Gateway Initiative (OSGi).

2. CONCEPT OF A CONSULTING MODULE IN ROOM AUTOMATION

When considering the exploitation phase of a building, the room is the basic optimisation object. The main user demands on comfort, ergonomy, psychology, physiology, simplicity, economy etc., should be taken into consideration. Examples of systems-integrating algorithms in the building automation are (Christen, et al., 1999):

- Window opening with actuator connected to the room automation with using the blocking of energy supply in incorrect controlling cases, i.e. „heating actuator off";
- Use of building heat accumulation ability;
- Reduction of the lighting costs through various regulating strategies depending on the outdoor light intensity and room occupancy.

From the technical point of view each building is a dynamic system offering therefore a certain potential for optimisation. The analysis of the influence of various measures (e.g. choice of the required value of room air temperature, of air conditioning and of the way of heating) on the energy consumption proved that, relatively large energy potential – up to 70 % - can be saved via reduced air conditioning (Hoh, 2002).

At the same time a building is a complex composed of a number of various service systems. The most processes within a building proceed in single rooms. As soon as the processes in each room are optimised the whole building reaches the state of optimal mode. Figure 1 illustrates the most important causal relations in room automation.

In the system integrating automation concept the following actuators/subsystems can change their conventional functions depending on the input values:

- Sunblinds can partly replace functions of actuators „lighting", „heating" and „cooling";

- Windows can partly replace functions of actuators „ventilation", „humidification", „heating" and „cooling";

- Ventilation can partly replace function of actuators "humidification", „heating" and „cooling".

Lighting and inner heat sources (computers, people) can also partly replace the function of „Heating" actuator. The following important rules for the so-called „function substitution" can be defined for certain situations:

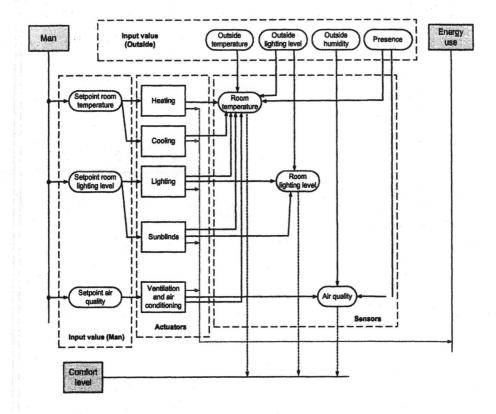

Fig. 1. The most important causal relations in room automation.

1. „Heating actuators" can be replaced by „sunblinds", „windows", „ventilation", „lighting" and „inner heat sources". In former three cases the free of charge transportation of the necessary heat energy from outside into inside is used;

2. „Cooling actuators" can be replaced by „sunblinds", „windows" and „ventilation", so the free of charge transportation of the needless heat energy from inside into outside;

3. „Ventilation" can be replaced by „windows", thus the free of charge transportation of the fresh air from outside into inside;

4. „Lighting" can be replaced by „sunblinds".

According to the figure 1 and the above-mentioned rules the situations leading to the highest energy wastes can be defined (Dementjev and Kabitzsch, 2003).

The resulting cost optimisation can be managed when the single actuators are known and the factors influencing the control loop can be detected. Each participant should intend the required state as the goal of all optimisation and consulting efforts. More important point is that the actual state of the room (e.g., human presence) should be known at any time (Olsen, 2002).

The tasks of the consulting module include:

1) Consulting function:

- Short- and long-term history accumulation;
- Recognition of typical states;
- Costs evaluation / energy consumption evaluation;
- Prognosis of situation development;
- User interaction.

2) Interface to management level:

- Connection with external data bases;
- Remote access for the user (visualisation).

3) Interface to field level:

- Field bus interface (LON);
- Logical interface to actual room state.

The system should inform the user how the change of the required value of the operative room temperature, CO_2 concentration and lighting level will effect the operational costs. Further estimated user activities are, for example, the scheduling of the sunblind operations.

It should be also noted that the ideas of such modules in room automation have been already expressed, e.g. "Office Assistant" (see (Hao Yan and Selker, 2000)), that helps the customer in the office in order to save staff's time; or "Networking Embedded Agents" (Huhns, 1999), that accumulate information about the state of the available household equipment and transmit orders in accordance with the day's schedule set in advance.

3. IMPLEMENTATION STEPS

An embedded PC was chosen as a hardware platform for the consulting module where further necessary software, among others frameworks and consulting algorithms, is installed.

An embedded server provides a software platform according to the OSGi standard that specifies a standard based on Java and comprises currently more than 80 members over the world.

Numerous research works appeared lately aiming to use such gateways (service-, home-, residential-, internet-gateways) for the room automation as a basis for various services and integration of hetero-geneous home networks (see Saito, *et al.*, 2000; Kastner and Leupold, 2001; Valtchev and Frankov, 2002; Gong, 2001; Soon Ju Kang, *et al.*, 2001)). The possible variants for realisation and application of service gateways are described in (Wacks, 2000). ISO/IEC JTC 1 (Joint Technical Committee 1), Information Technology, SC 25 (Subcommittee 25), Interconnection of information technology equipment, WG1 (Working Group 1), Home electronic system are also concerned with the "Home Gateways" issue.

The OSGi framework (Figure 2) uses the concept of a Java virtual machine that provides approaching of required flexibility for the multitude of services and requirements.

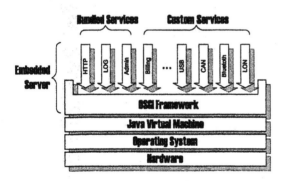

Fig. 2. OSGi Software Architecture

The OSGi aim is the definition and distribution of open specifications allowing the supply of various services between external and local networks to the final devices.

The previous proprietary solutions implied that the hardware of all levels (devices, PLC, HMI) should be produced by the same single supplier. That presents a considerable limitation because the designer cannot choose the best combination of hardware parts for his solution out of various suppliers. He has to stick to one supplier instead that offers him only one compromising solution. An important task is also

hard- and software update. The major problem for the hardware modernisation is the development of new controlling algorithms and user's interfaces for the corresponding hardware. While updating software the whole system should be stopped that demands not only time but additional costs as well (Hackbrath, 2001).

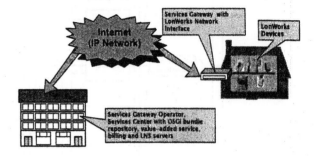

Fig.3. OSGi Architecture and LONWORKS ® Technology

By development of the consulting module was used LONWORKS Bundle Deployment Kit of Echelon Corporation (specially for inexpensive solutions and based on the LNS management software installed in the central office of the gateway's operator, see Figure 3). The Bundle is compatible with every framework corresponding to the 1.0 OSGi specification including Java embedded server of Sun.

Figure 4 illustrates consulting module concept basing on the OSGi architecture and serving as the interface between field- and management levels.

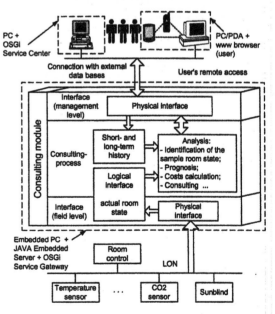

Fig. 4. Consulting Module Architecture (Kabitzsch, *et al.*, 2002b).

Consulting algorithms are developed on the basis of fuzzy logic (module of setting and controlling of comfort level) and neural networks (prognosis of changes in the level of comfort and energy consumption) because creation of a conventional mathematical model for the real process can be very expensive, and the most of mathematical models contain a lot of simplifications and linearization requiring step-by-step optimisation. Thus the „classical" way is mostly not realisable or too expensive for practical applications. As the simplifications and linearization of the model influence the regulators, consulting quality depends to large extent on the simplifications chosen partly accidentally. Such tasks, e.g. the prognosis of the user's behaviour, can be better solved using fuzzy and neural algorithms as well.

For checking of the efficiency of the developed algorithms was used a model showing the basic physical processes in a room. Results of the first simulations made in MATLAB proved that the offered algorithms let save up to 32% of consumed energy by heating (simulation of two weeks in winter with and without consulting module) and up to 36% of consumed energy by cooling (similar simulation for summer).

For solving of the task of the prognosis of changes in the level of comfort and energy consumption there was developed a neural network model showing the basic physical processes in a room. For training of the network there was used the data of the simulation of the room model in MATLAB ® Simulink. Neural network imitates the system of 14 inputs (actuators state, outer environment parameters etc.) and 7 outputs (comfort parameters: room temperature, light, air humidity and CO_2 and energy consumption for heating, lighting and cooling).

Figure 5 presents the neural network simulation results for room temperature. The maximum relative error by simulation is 3,53 % and the average relative error is 0,96 %.

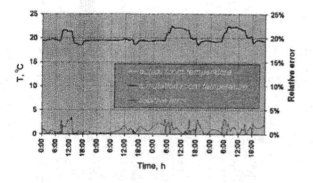

Fig. 5. Neural network simulation results for room temperature

For testing of the possibilities of realisation of consulting algorithm was developed a physical room model (Fig. 6) equipped with all the necessary sensors for defining of the level of comfort (outer and inner temperature, relative humidity and lighting, air quality), with actuators (sunblinds, ventilators, lamps, heating and cooling elements) and controllers (temperature, ventilation regulators, sun-blinds and lighting controller).

Fig. 6. Room model for testing of the algorithms

4. CONCLUSION

Consulting algorithms were developed on the basis of fuzzy logic and neural networks with corresponding development- and simulation-environment and the first simulation results already exist.

To test the algorithm was made the room model in MATLAB ® Simulink (representing the most important physical parameters of a room) and the physical room model.

While testing the proposed service models an embedded server with OSGi frameworks was installed and the possibilities of remote access to LONWORKS ® network per Internet were proved.

Further the consulting algorithms will be implemented with Java and ported in an embedded PC.

5. ACKNOWLEDGEMENT

The project the present report is based on was supported by the German Federal Ministry of Education and Research under the registration number 01HW0131. The authors bear all the responsibility for contents.

REFERENCES

Christen, M., K. Friedrichs and E. Jund (1999). *GNI - Handbuch der Raumautomation. Gebäude-technik mit Standardsystemen.* VDE Verlag, Berlin Offenbach

Corcoran, P.M., F. Papal and A. Zoldi (1998). User interface technologies for home appliances and networks. In: *IEEE Transactions on Consumer Electronics.* Vol. 3. pp. 679 – 685.

Dementjev, A., Kabitzsch, K. (2003). A Consulting Module in Room Automation. In: *Proc. IFAC Workshop Modelling and Analysis of Logic*

Controlled Dynamic Systems, Irkutsk, Russia. pp. 41 – 46.

Gong, L. (2001). A software architecture for open service gateways. In: *IEEE Internet Computing*. Vol. 1. pp. 64 – 70.

Hackbrath, K. (2001). Architekturen für Industrienetzwerke. In: *SPS-Magazin*. Vol. 10. pp. 112 – 113.

Hao Yan and T. Selker (2000). Context-aware office assistant. In: *2000 International Conference on Intelligent User Interfaces*. New Orleans, USA. pp. 276 – 279.

Hoh, A. (2002). Veränderung des Jahresheizenergie-bedarfs von Büroräumen durch Nutzereingriff – Simulationsergebnisse. Bericht. TU Dresden, Institut für Thermodynamik und Technische Gebäudeausrüstung.

Huhns, M.N. (1999). Networking embedded agents. In: *IEEE Internet Computing*. Vol.1. pp. 91 – 93.

Kabitzsch, K., D. Dietrich and G. Pratl (2002a). *LonWorks Gewerkeübergreifende Systeme*. VDE Verlag. Berlin Offenbach.

Kabitzsch, K., R. Schach and G. Knabe (Eds.) (2002b). *Dokumente des Workshops „Brücken zwischen Gebäudeautomation und Facility Management"*. TU Dresden

Kastner, W. and M. Leupold (2001). Discovering Internet Services: Integrating Intelligent Home Automation Systems to Plug and Play Networks. In: *Lecture notes in computer science*. Vol. 2060. pp. 67 – 78.

Kuntze, H.-B. and G. Nirschl (1998). Prototypische Realisierung einer benutzerfreundlichen fuzzy-basierten Raumklimaregelung. In: *Jahresbericht Fraunhofer-Institut Informations- und Datenverarbeitung IITB*. pp. 54 – 55.

Laukner, G. and G. Knabe (2001). Bedarfsgeführte Einzelraumregelung für Heizung und dezentrale Lüftung. In: *Dokumente des Forschungsprojektes*. Fraunhofer-Institut für Verkehrs- und Infrastruktursysteme IVI; TU Dresden, Institut für Thermodynamik und Technische Gebäudeausrüstung. URL: http://www.ivi.fhg.de/frames/german/fields/index.html

Olsen, J. (2002). Web-basierendes Informationsmanagement in der Gebäudeautomation. In: LNO-Brief. Vol. 25. pp. 18 – 19.

Saito, T., I. Tomoda, Y. Takabatake, J. Arni and K. Teramoto (2000). Home Gateway Architecture and its Implementation. In: *IEEE Transactions on Consumer Electronics*. Vol. 4. pp. 1161 – 1166.

Schach, R., K. Kabitzsch and G. Knabe (2001). Wissensintensives Gebäudemanagement. Vorhabenbeschreibung. BMBF–Fördermaßnahme "Wissensintensive Dienstleistungen". TU Dresden

Soon Ju Kang, Jun Ho Park and Sung Ho Park (2001). Room-Bridge: Vertically Configurable Network Architecture and Real-Time Middleware for Interoperability between Ubiquitous Consumer Devices in the Home. In: *Lecture notes in computer science*. Vol. 2218. pp. 232 – 251.

Tamarit, C. and G. Russ (2002). Unification of Perception Sources for Perceptive Awareness Automatic Systems. In: *Proc. 6th IEEE AFRICON Conference, George, South Africa*. Vol. 1. pp. 283 – 286.

Tarrini, L., R. Bianchi Bandinelli, V. Miori and G. Bertini (2002). Remote Control of Home Automation Systems with Mobile Devices. In: *Lecture notes in computer science*. Vol. 2411. pp. 364 – 368.

Valtchev, D. and I. Frankov (2002). Service Gateway Architecture for a Smart Home. In: *IEEE Communications Magazine*. Vol. 4. pp. 126 – 132.

Wacks, K. (2000). Guiding the Home Systems Industry to Smarter Interfaces. In: *ISO Bulletin*. Vol. 12. pp. 4 – 11.

Copyright © IFAC Telematics Applications in Automation
and Robotics, Espoo, Finland, 2004

ELSEVIER

IFAC
PUBLICATIONS
www.elsevier.com/locate/ifac

OPEN SERVICE ARCHITECTURES FOR ENVIRONMENTAL MONITORING AND INDUSTRIAL AUTOMATION

Seppo Sierla, Jukka Peltola, Kari Koskinen, Teemu Tommila

Helsinki University of Technology
Information and Computer Systems in Automation
P.O.Box 5500, FIN-02015 HUT, Finland
[Seppo.Sierla, Jukka.Peltola, Kari.O.Koskinen]@hut.fi
VTT Industrial Systems
Production Systems
P.O.Box 1301, FIN-02044 VTT, Finland
Teemu.Tommila@vtt.fi

Abstract: This paper discusses globally distributed telematics applications in the environmental monitoring and industrial automation application areas. The requirements on the necessary distributed computing infrastructure are described. Based on these requirements, the usefulness of component-based software architectures and open service architectures is discussed, and the applicability of the Open Services Gateway initiative (OSGi) in these applications is evaluated. *Copyright © 2004 IFAC*

Keywords: Communication environments, Communication systems, Distributed computer control systems, Remote control, Telematics

1 INTRODUCTION

In recent years, there has been a clear trend away from stand-alone equipment and process deliveries and towards global services for these products. A distributed computing infrastructure is essential for any remote service such as monitoring or diagnostics. There is much diversity among such applications, so the requirements for the supporting IT systems vary greatly. This paper discusses globally distributed telematics applications in the environmental monitoring and automation industries. In our previous research (OHJAAVA-2 2003), an open component-based software architecture has been identified as an essential element in the next generation of distributed control systems (Tommila, et al. 2001). Now this architecture is being extended for use in environmental monitoring and remote industrial automation, and suitable standards and technologies are being evaluated (OHJAAVA-3 2003).

Many technologies provide support for global communications, versatile local area communication, software lifecycle management, run-time platforms for small embedded devices, openness in a multi-vendor environment or changing business models. The strengths of some promising technologies are summarized in this paper, but it is clear that this long list of disparate yet necessary requirements is not easily satisfied, although many solutions are strong in some of these areas. The OSGi (Open Services Gateway initiative) is chosen for detailed evaluation, because it promises support for all of these requirements.

The OSGi designers have carefully avoided focusing on a certain application area or network technologies. However, they have considered residential gateways, mobile phones and vehicles as typical applications (Marples and Kriens, 2001) (OSGi 2003). Several articles focus on incorporating networking technologies that are suitable for homes and discuss advanced service discovery features in such environments (Dobrev, et al., 2002) (Wils, et al., 2002). Most of the work is not directly applicable to industrial applications, although (Campos, et al., 2002) demonstrates the usefulness of OSGi for managing a fleet of vehicles. This paper discusses the integration of OSGi with such architectural models and communication mechanisms that satisfy the requirements of industrial applications (OHJAAVA-2 2003).

2 THE APPLICATION SCENARIO

The main elements in the globally distributed telematics systems that are discussed in this paper are sites where the industrial processing takes place. The experts with the necessary know-how for managing the sites are located in service centers and other offices that might be on another continent. The sites are connected to a WAN (Wide-Area Network) by some gateway, and usually each site communicates with a central server system or single node such as the laptop of a diagnostics expert. In the simplest case, the site contains only one node, so this is similar to a vehicle (a common OSGi scenario). However, the site might consist of collaborating nodes, which communicate with rich communication mechanisms suitable for industrial applications; the gateway to the site would then offer suitable services which are provided transparently by the community of nodes within the site. Typical services are diagnostics, monitoring, parameter-based control and software updating.

The above scenario is typically found in environmental monitoring and industrial automation applications. The automation applications are extended services to products and systems that operate at sites around the world. The site in this case is a part of a factory where the main automation system might been supplied by another vendor. For example, remote diagnostics, monitoring, predictive maintenance and optimization of a machine or process are a great source of business opportunities in an industry where the demand for plain machines and automation systems has been saturated (Ketonen 2001). Another example of a site is crushers at gravel pits, where no human operator is present. There might be several moving machines on the site, which should collaborate and provide some service access point for remote users.

3 CASE: ENVIRONMENTAL MONITORING AND INDUSTRIAL AUTOMATION

Weather monitoring stations are installed to urban as well as remote locations, such as forests or jungles. The quality of the local publicly available communication infrastructure varies considerably. The station has sensors connected to it for measuring such variables as temperature, humidity and wind speed. The station's main responsibilities are sending a few weather reports per hour or alarms and notifications whenever the need arises. It is also desirable that measurement data can be queried and parameters can be adjusted. What measurements are made and what software functionality is needed will vary among stations. With some clients, significant cost savings are possible if software components can be updated remotely without sending a technician on-site.

Weather monitoring stations have some special requirements for the IT infrastructure. A station typically only communicates with a central server and a database system. The most affordable, reasonably reliable communication technology is used; usually GSM/GPRS or local telephone lines are chosen. Since there are no tight real-time requirements on the data transfer, most of the effort is directed at minimizing the cost of the communication. Therefore, messages are coded into a compact, mostly binary, format, and the bandwidth consumption of popular Internet protocols such as SOAP is too high. Similarly, the hardware should be cheap and robust especially since power consumption must be limited and maintenance visits avoided. There might only be a few MB of memory, and usually no hard disk is available.

The situation is more complex at an airport with several environmental monitoring units connected to a LAN, at a mineral processing site with a community of collaborating, moving machines or a factory where a vendor is responsible for some aspect of the operation of a process. Support is needed for multi-vendor environments, rich communication mechanisms, reliable real-time Quality of Service and component lifecycle management with minimal or zero downtime. The next section describes the requirements for a software architecture that should satisfy the needs of any of these cases.

4 COMPONENT BASED SOFTWARE ARCHITECTURE

The key issues for industrial telematics applications can be summarized as follows:

- modular products and platforms that enable easy aggregation and parameterisation according to customer's needs during the design phase
- support for remote diagnostics, software updates and value-added services
- light-weight solutions for limited computing resources and communication bandwidth
- intelligent communication middleware that optimises communication costs and adapts to the often unreliable communication infrastructure
- use of mainstream technologies that provide for easy integration and long-term support
- possibility to include local distribution services, peer-to-peer (P2P) functionality, plug&play configuration in the future

In a natural way, these requirements lead to a component based software architecture that provides an execution environment both for the components of the system software and the application (Figure 1).

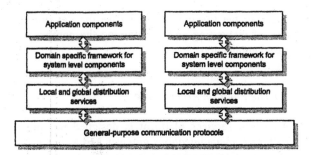

Figure 1. Architecture layers of a generic IT infrastructure (Tommila, et al., 2001)

The components distributed to local small-scale nodes and connected to global servers should have a rich set of interaction mechanisms according to the needs of the application area. In our case these include data distribution, event notification and services. Intensive communication such as data distribution might take place locally within the site, for example among a group of moving machines. The gateway would provide access to its services with such mechanisms that are suitable for use over a WAN. Figure 2 illustrates the possible communication scenarios for industrial automation; these mechanisms are fully adequate for weather monitoring as well.

This situation is not very unique in the industry, and several approaches in this direction have been suggested, including for example (Enterprise) Java Beans, application servers, etc. However, most of them have been either designed for local (low-cost) applications or for distributed applications based on high-end computing platforms (such as Enterprise Java Beans). The next section describes OSGi as a possible solution for our needs.

5 OPEN SERVICES GATEWAY INITIATIVE

Several promising technologies exist, which can be used to construct comprehensive telematics infrastructures. Some cell phone manufacturers for example have embedded J2ME IMP platform in small devices with GPRS/GSM and IO capabilities. However, these m2m (machine to machine)

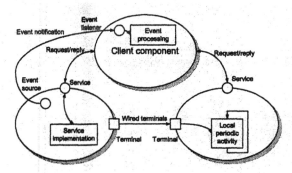

Figure 2. Interaction mechanisms of distributed components.

platforms are capable of hosting only a single java application ('imlet') at a time and they lack support for running and managing modular component based applications. Management of a plentitude of platforms may become a daunting task and requires advanced remote management applications. In such case it would be desirable for the platform to have standard management interfaces, like Java Management Extension (JMX), to avoid being tied up to proprietary management tools, such as the Nokia's m2m gateway. OSGi Service Platform is a framework built on a java platform, which allow dynamic management and extensibility of applications, standard interaction mechanisms between components, remote management and many other useful services for telematics applications.

5.1 The need for OSGi

OSGi's mission is to "create open specifications for the network delivery of managed services to local networks and devices" (OSGi 2003). It is important that no particular services or application areas are singled out; rather, the reference architecture is general enough to be used with a great variety of applications and business models. As networks and especially the Internet become increasingly ubiquitous, remote monitoring, maintenance and diagnostics services are being developed for all kinds of devices. However, individual companies rarely have time to design coherent architectures, so adapting solutions to different clients and circumstances usually requires much tailoring. Resorting to proprietary non-standard solutions naturally creates serious interoperability and integration problems.

5.2 The OSGi Reference Architecture

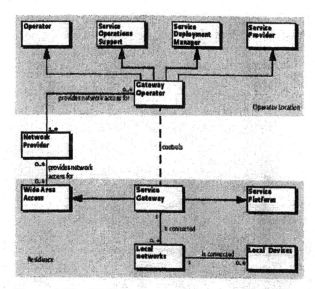

Figure 3 The OSGi Reference Architecture (OSGi 2003)

Figure 3 illustrates the primary reference architecture, where a number of globally distributed service platforms are being managed (OSGi 2003). The platforms contain an implementation of the OSGi framework and run a set of bundles, which are components that provide the services. The platform is hosted on a Service Platform Server (SPS), which is typically assembled from the cheapest hardware components that can run the application. The platform is connected to devices using some local network or I/O technology. The Service Gateway incorporates the platform and also has WAN access functionality.

The upper part of Figure 3 shows the actors that are involved in controlling the service gateways (OSGi 2003). The service provider is the organization that has developed the services for the platforms. Related to this is the Service Operations Support (SOS) at some central server, which cooperates with the software on the platforms. The Service Deployment Manager (SDM) installs and updates the services onto the platforms, and the Operator controls which SDMs may access a particular Service Platform. Finally, the Network Provider manages connectivity to the WAN, which usually is Internet.

5.3 Applying the Architecture to Weather Monitoring

A weather monitoring station corresponds to a Service Gateway. The platform would ideally be a small OSGi implementation, perhaps on top of Java 2 Micro Edition and CLDC. Typical Service Applications could be the generation and sending of weather reports or the management of parameters. The platform is hosted on a Service Platform Server (SPS), which is built of cheap and robust hardware that has been designed for rugged environmental conditions. The platform is connected to the sensor devices using simple I/O. WAN access functionality is usually GPRS or modem over a local phone line. The OSGi model does not assume that the WAN is the Internet or that there is a continuous connection; neither assumption would be valid in our case.

The reference architecture in Figure 3 accommodates external service providers. The providers of weather monitoring systems and services might belong to the same or different organizations and their relationships will depend very much on the circumstances. Figure 3 accommodates the situation where all of the illustrated roles are performed by different organizations. In practice, a single organization will often be responsible for several roles. An open service architecture is equally useful for coordinating responsibilities within the organization. Should it be necessary, at some point, to procure software or services from external providers, an open architecture will facilitate the sharing of responsibilities and integration of software. Whatever the business model, the

interoperable OSGi tools will ease the developers' task of integrating and maintaining the component-based application.

5.4 The Platform Architecture

An OSGi services platform is designed to operate both as a *communications gateway* and as a *services gateway*. Communications is passed between locally networked devices (LAN) and a wide area network (WAN). Remotely managed applications provide services to the users e.g. in vehicles, at homes or at industrial locations. In case of remote weather monitoring applications, the user is a central monitoring application or possibly another monitoring station nearby both residing behind a WAN connection. A central monitoring application gathers reports sent by a multitude of monitoring stations, stores and processes the data and makes it available to other applications through database connectivity.

The key aspect in OSGi service platform architecture is a bundle. A bundle is a remotely managed software component, which can interact with other bundles through standardized APIs and mechanisms of the OSGi framework. A bundle is accessed by the surrounding framework through an interface, which is separate from the bundle's implementation. This separation allows changing the implementation without changing the environment and other bundles. Thus, a bundle can have multiple implementations to adapt to different resource configurations or communication protocols. The OSGi services framework separates the local area sensor access mechanisms, the wide area access and the service applications from each other. This brings flexibility to the selection and design of these components and allows for a robust application structure.

The service developer builds his application into application bundles and relies on the bundle interaction mechanisms defined by OSGi. The application may use services provided by the platform's Service Bundles, e.g. Http Service and services provided by Extension Bundles, typically device drivers and protocols. The interaction mechanisms defined by OSGi are plain method calls to service methods registered to the framework by bundles. A richer set of communication mechanisms suited to the needs of industrial automation or weather monitoring should be available. This could be implemented as a sort of 'middleware bundle' between application bundles and communication resources. Similar development needs have been discovered for example in (Choonhwa, et al., 2003).

Figure 4 shows a logical picture of what kind of components are needed in the platform. The top and bottom rows present some basic communications

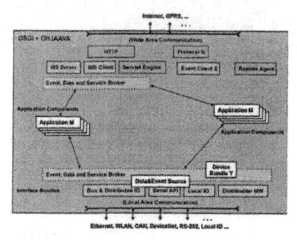

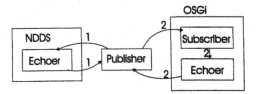

Figure 4 OSGi Service Platform running several application components

Figure 5 Test arrangements

service bundles, which provide access to network and IO resources. The *Event, Data, and Service Broker* is an abstraction of a middleware service, which implements the communication mechanisms needed by the application components. The need for such a layer was identified in OHJAAVA research and the Real-Time Publish-Subscribe (RTPS) standard was evaluated as an interesting solution for our communication needs for industrial applications; RTPS supports the mechanisms in Figure 2 (RTI 2003; Sierla, et al., 2003). The following chapter presents a test setup, where the distribution service (RTPS) has been loaded onto the service platform as an OSGi bundle. It can be regarded as a single bundle version of the two components in Figure 4: *Event, Data, and Service Broker* and the *Distribution MW*.

5.5 Local Distribution Experiments

The experimental work is based on a scenario where components at a site need to collaborate using rich communication mechanisms. Data distribution for cyclic measurement or control activities must proceed in real-time, but extended services such as diagnostics bundles on the OSGi platform might also want to receive this data. During more or less unexpected situations, there will be a need for transmitting events, alarms and notifications to all interested parties, including bundles on the OSGi framework as well as other components. Implementing such communications using the OSGi framework's mechanisms for importing services from other bundles would require much work and could not support the automatic updating of communication links as the distributed application is reconfigured. (Choonhwa, et al., 2003) describes event services based on the producer-consumer model for home automation, but appropriate communication technologies for industrial applications have not yet been integrated as commercially available bundles. The bundles in our demonstration environment use the Real-Time Publish-Subscribe (RTPS) middleware, which has

been identified as a promising solution for the needs of new industrial applications (OHJAAVA-2 2003). The main purpose of these tests is to determine the interoperability of the NDDS implementation of RTPS with an OSGi platform, since these are independently developed techniques, but some performance results are also presented.

Figure 5 illustrates the two test scenarios. In the first test (with data flows labelled 1 in the figure), 10000 measurements were sent using a NDDS publisher with a period of 10ms. The receiver was a stand-alone NDDS component, which echoed a response to the sender, so the latencies here are round-trip times. The measurement in this case was an integer sequence number and double timestamp. The test was run on a 2GHz, 512MB Windows XP machine. In the second scenario (with messages labelled 2), the same echoer was registered as 2 OSGi bundles. One of these subscribed to the published messages and imported the other bundle, which provided a service for sending the echo.

Results for scenario 1 are labelled as 'NDDS' in Figure 6, which shows statistics for round trip times; the results for the second scenario are labelled 'OSGi'. The statistics show that the overhead of running NDDS on the OSGi framework as well as the communication delay between two bundles is insignificant, when compared to the lean NDDS real-time middleware operating on top of UDP sockets. The minimum, average, maximum and standard deviation give a good approximation of best, average and worst case performance as well as jitter; the best way to improve the worst case performance and jitter is to move to a real-time OS.

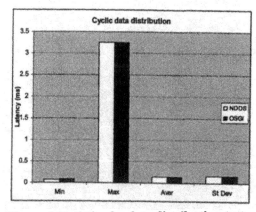

Figure 6 Statistics for data distribution test

Our previous experiences with NDDS indicate that the overhead from a 100Mbps Ethernet is minor (OHJAAVA-2). Running any tests over the undeterministic Internet would obscure the performance of the OSGi platform, which is the object of this test; much research has already been carried out on network performance, which is outside of the scope of this paper.

Figure 7 shows results for a similar test when bursts of 5 events are sent every 30ms, for a total of 10000 messages; such bursts are common in exceptional situations. The test arrangements as well as the message structure were the same as in the previous test. Although the latencies are slightly higher, no problems were encountered. The higher maximum for the OSGi echoer is presumably caused by the Windows scheduling. In summary, these initial results suggest that the use of OSGi is very feasible in industrial applications with real-time data distribution and event communications. However, if the OSGi bundles are involved in real-time control, the underlying Java Virtual Machine should implement a real-time Java specification.

6. CONCLUSIONS AND FURTHER WORK

Open service architectures and component-based software architectures provide a rich, flexible and robust framework for application developers in the environmental monitoring and automation industries. When integrated with communication technologies for industrial processing, OSGi is an interesting solution for building extended services such as remote monitoring and diagnostics. If a future implementation supports real-time Java, OSGi can even be used as a platform for executing and managing the entire component based industrial application.

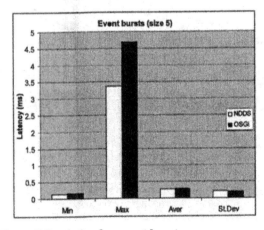

Figure 7 Statistics for event bursts

7. REFERENCES

Campos F.T., W.N. Mills, M.L. Graves (2002). A Reference Architecture for Remote Diagnostics and Prognostics Applications. In: *AUTOTESTCON Proceedings*, 842-853. IEEE.

Choonhwa L, D. Nordstedt, S. Helal (2003). Enabling Smart Spaces with OSGi. In: IEEE Pervasive Computing, July-September 2003, 89-94.

Dobrev P., D. Famolari., C Kurzke., B Miller (2002). Device and Service Discovery in Home Networks Via OSGi. In: *IEEE Communications Magazine*, **August 2002**, 86-92.

Ketonen M., A. Marttinen (2001) Real-Time Control Performance Monitoring as a Tool for Remote Maintenance Service. In: *1st IFAC Conference on Telematics Applications in Automation and Robotics*, **TA 2001**, 271-276. IFAC, Weingarten, Germany.

Marples D., P. Kriens (2001). The Open Services Gateway Initiative: An Introductory Overview. In: *IEEE Communications Magazine*, **December 2001**, 110-114.

OHJAAVA-2, Modern Distribution Solutions in Open Control Systems, http://www.automationit.hut.fi/tutkimus/docum ents/Ohjaava/eohjaava-2.htm, 6.10.2003

OHJAAVA-3, Geographically Distributed Monitoring and Control System, http://www.automationit.hut.fi/tutkimus/docum ents/Ohjaava/eohjaava-3.htm, 6.10.2003

OSGi: OSGi Service Platform Release 3. March 2003, http://www.osgi.org

RTI: Real Time Innovations Inc. http://www.rti.com/ 1.10.2003

Sierla S., J. Peltola, K. Koskinen (2003). Evaluation of a Real-Time Distribution Service. In: *SoftSympo 03 Proceedings*, September 2003, Helsinki, Finland.

Tommila T., J. Peltoniemi, J. Peltola, S. Sierla, K. Koskinen (2003). Uudet hajautusratkaisut avoimissa automaatiojärjestelmissä. In: *Automaatiopäivät 2003 Proceedings*, September 2003, Helsinki, Finland.

Tommila T., O Ventä, K Koskinen (2001). Next Generation Industrial Automation – Needs and Opportunities. In: *VTT Automation, Automation Technology Review*, **2001**, 34 – 41 http://www.vtt.fi/tuo/projektit/ohjaava/ohjaava1 /atr_2001.pdf.

Wils A., F. Matthijs, Y. Berbers, T. Holvoet (2002). Device Discovery Via Residential Gateways. In: *IEEE Transactions on Consumer Electronics*, **Vol 48 No. 3, August 2002**, 478-483

Copyright © IFAC Telematics Applications in Automation and Robotics, Espoo, Finland, 2004

ELSEVIER

IFAC
PUBLICATIONS
www.elsevier.com/locate/ifac

MODEL-BASED APPROACH TO CHARACTERIZATION OF DIFFUSION PROCESSES VIA DISTRIBUTED CONTROL OF ACTUATED SENSOR NETWORKS

Kevin L. Moore and YangQuan Chen

Center for Self-Organizing and Intelligent Systems
Utah State University
4160 Old Main Hill,Logan, Utah 84322
moorek@ece.usu.edu

Abstract: This paper presents a model-based distributed control framework for the problem of characterizing diffusion processes. Such systems arise in a wide variety of applications. We propose the use of multiple, robotic sensors that can be tasked to collect measurements of the process outputs from variable spatial locations. The notion is that a central controller uses the process model to compute a prediction of the process outputs. This prediction is then used to prescribe sampling locations, to which the sensors in a distributed sensor network are commanded. Samples obtained by the sensors are then used as constraints to correct the model-based prediction of the process outputs. The cycle of predicting, sampling, and prediction correction is repeated until the predictions converge. The idea of actuated actuator networks is also introduced. A control-theoretic formulation of the problem is presented that highlights a number of theoretical problems to be solved. *Copyright © 2004 IFAC*

Keywords: Autonomous mobile robots, sensor networks, distributed control.

1. INTRODUCTION

Through the course of history many technologies have developed to the point of becoming ubiquitous. For example, the use of electric motors is today so pervasive that they are taken completely for granted by most people. Similarly, microprocessors and other digital hardware, such as FPGAs, are now found in every conceivable application – from washing machines to cell phones – to such an extent that it will soon be hard to imagine life without them.

Today, a new "technology" is developing that also has the promise of ubiquity. Due to a combination of advances in biology, electronics, nanotechnology, wireless communications, computing, networking, and robotics, it is now possible to:

1. Design *advanced sensors and sensor systems* that can be used to measure an increasingly wide range of variables of interest, from explosives and toxic, biological, and chemical agents, via the design and synthesis of functionalized receptors and materials; to the detection of objects and events, via advances in multi-spectral imaging systems, data fusion, and pattern recognition techniques; to sensors for physical measurements, such as force, pressure, acceleration, flow, vibration, temperature, humidity, and physical variables, such as mass, magnetic, optical, ultrasonic, chromatographic, and others.

2. Use *wireless communications*, or telemetry, to effectively communicate sensor data from a distance. Sensors can now be deployed in fixed locations or via mobile robots, without the constraint of wires, to operate in extreme environments, possibly autonomously, allowing humans to observe data while staying out of harm's way.

3. Build *networks of sensors*, using wireless communications and computer networking technology, which can provide the capability to obtain spatially-distributed measurements from low-power sensors that communicate and relay information between each other. Such sensor networks can be homogeneous, using a single type of sensor, or heterogeneous, employing multi-modal sensors, whose outputs are combined and interpreted through data fusion algorithms.

4. Develop *reconfigurable, or adaptable, networks of distributed sensors* by providing mobility or actuation to the individual sensors in the network. For example, if each sensor was mounted on an autonomous robot, the network could perhaps self-organize, moving sensors to the appropriate locations needed to ensure adequate coverage.

Current research on these topics is leading to the promise of a new ubiquitous technology – distributed sensor networks – where "data about everything is available everywhere." However, with this promise comes a related challenge: how to interpret and use this data and for *what purpose*. Moreover, with a mobility platform, the networked sensors and actuators can be actively moved according to a high-level task or mission. There is a strong need for system level, mission-centered research for mobile actuator-sensor networks. In the remainder of this paper we propose such research, related to the interpretation and use of data in distributed sensor networks, with a focus on coordination strategies for networks of mobile sensors. We first discuss some issues related to such networks and then present a specific application scenario that is the basis for our main proposal: a model-based approach to sensor coordination. Next we formulate a general control-theoretic problem applicable to a large class of mobile sensor application scenarios. We continue by introducing one final consideration: the notion of a network of mobile actuators combined with a network of mobile sensors.

2. DISTRIBUTED SENSOR NETWORKS

Sensor networks are drawing increased attention from research communities, industry sectors and government agencies. As stated in (National Science Foundation, 2003), sensor networks will *"have significant impact on a broad range of applications relating to national security, health care, the environment, energy, food safety, and manufacturing. The convergence of the Internet, communications, and information technologies with techniques for miniaturization has placed sensor technology at the threshold of a period of major growth."* Recent surveys on sensor networks (Hairong, *et al.*, 2001; Akyildiz, *et al.*, 2002) also indicate the importance of sensor networks research. Many on-going efforts are focused on various specific issues in sensor networks such as the sensor structures (Abdelzaher, *et al.*, 2003; Yang and Sikdar, 2003), communication (Akyildiz, *et al.*, 2002), data processing and sensor fusion methods (Hairong, *et al.*, 2001; Kumar, *et al.*, 2002), sensor deployment and localization (Hairong, *et al.*, 2001; Kumar, *et al.*, 2002; Wang, *et al.*, 2003), calibration (Whitehouse and Culler, 2002; Bychkovskiy, *et al.*, 2003), etc. However, from a dynamic systems and control point of view, the sensor networks should be part of a complete system with a specific mission defined. Recently, a habitat monitoring task was proposed as an "application driver" for wireless communications technology based sensor networks (Cerpa *et al.*, 2001), although this is still an open-loop system simply for monitoring purposes. In (Sukhatme *et al.*, 2000) a future research effort was proposed to combine distributed sensing, robotic sampling, and offline analysis for *in situ* marine monitoring purposes, where the loop is closed using underwater robots that carry networked sensors and are deployed according to the sensed environment. This system is not real-time feedback-controlled due to the offline analysis. So far, there is no such real-time closed-loop

distributed feedback control system involving networked actuators and sensors (Haenggi, 2002).

2.1 A Motivating Example.

To motivate our subsequent development, consider the sequence of illustrations given in Figs. 1-3. Fig. 1 depicts a situation where some type of biological or chemical contamination has occurred. As a result, there is a developing plume of dangerous or toxic material. Of course, the exact development of the plume will depend upon the prevailing weather conditions as well as the surrounding geography. There are sophisticated models that can be used to predict the development of the plume. Such models are based on partial differential equations that include diffusion and transport phenomena effects as well as forcing functions such as the prevailing weather conditions and boundary conditions and constraint equations defined by the surrounding geography. As shown in Fig. 1, such models can be used to build (an imperfect) prediction of the plume boundary. Next, assume that we have actuated sensors that can measure the concentration of the contaminant. These sensors are deployed in a swarm as a mobile sensor network. Their motion is coordinated so as distribute them around the predicted plume boundary as shown in Fig. 2. Existing work on swarm behavior often uses potential or attractor fields to coordinate swarm motion. The distinction in our approach is that the attractor field we use to coordinate the swarm motion is derived from a physics-driven model of the task. Note that we assume the individual actuated sensors are an appropriate type of robot, in this example a flyer, and would have autonomous guidance and navigation (with waypoints defined by the predicted plume boundary). All the actuated sensors would be able, in this example, to communicate wirelessly with each other using some type of protocol to form an *ad hoc* network and with some presumed base station (where the actual plume prediction takes place). Of course, as shown in Fig. 2, the plume continues to grow as the mobile sensor network collects its samples.

Fig. 1. Initiation of a diffusion process in an urban environment and prediction of the plume.

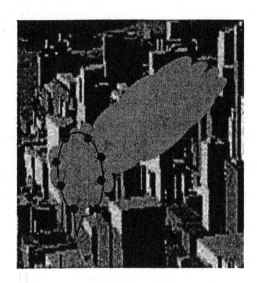

Fig. 2. A mobile sensor network collects samples from the predicted plume location, while the plume continues to grow.

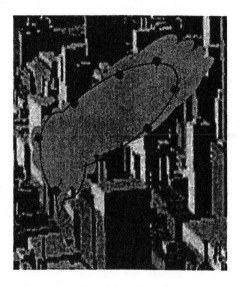

Fig. 3. New prediction of the diffusion process states and collection of data at new locations.

But, as shown in Fig. 3, using the physics-based model and data about prevailing weather and geography, but now with the additional information of actual concentrations gathered from the swarm of actuated sensors, a new prediction of the plume boundary can be computed and, after the new predicted plume boundary is computed, the swarm's attractor function is modified and the network of actuated sensors autonomously deploys to a uniform distribution around the new predicted boundary, where they collect new samples, which are relayed to the base station, where a new prediction is made and then the process continues. As more and more samples are collected, the predictions would ideally converge to the true plume location.

2.2 Issues Related to Distributed Sensor Networks.

Motivated by the application scenario above, we see that the elements of a distributed, mobile, robotic sensor net should include:

1. A *sensor* relevant to the application at hand. For some classes of measurement problems, commercial off-the-shelf (COTS) sensors already exist, though work continues to make them smaller, more robust, and lower-power. In other applications, especially chemical-biological, sensor development is currently a subject of intense activity. For our purposes we suppose COTS sensors are available.

2. Some type of (typically wireless) *communication* capability. This could be radio frequency, microwave, or cellular communications, depending on the power and range requirements. At this time perfect COTS solutions do not exist, especially if the sensors might be deployed indoors and/or outdoors. However, a number of research teams and companies are actively pursuing development of wireless communication technologies, so again, for our purposes, we assume an appropriate communication capability is available.

3. Suitable *communication and data flow control protocols*. A set of sensors with communication hardware only become a network when they can communicate according to some standard. The issue is to ensure that available network resources are allocated to optimally control the data flow between nodes (i.e., sensors). Again, the development of such protocols has been an active area in recent year, particularly due to the emergence of computer networking solutions (e.g., the "mote" concept, the 802.11b standard, the FieldBus and CAN bus standards, *ad hoc* and infrastructure network protocols, etc.). For now we assume suitable protocols, but note that as one moves to consider integration of control efforts with distributed sensors, the existing protocols are inadequate and research may be required to achieve our ultimate vision of a network of mobile actuators combined with a network of mobile sensors.

4. Algorithms for *sensor placement and data interpretation and use*. Such algorithms are necessarily application-dependent. Generally, domain-specific expertise is adequate to address the issue of data interpretation. However, data interpretation is also related to a less well-developed topic: sensor placement to achieve a given task. For example, in the scenario of Figs. 1-3 one can ask: how are the mobile sensors optimally located? We submit that to date such questions have been handled in an *ad hoc* fashion. Below we establish a systematic way to address this issue for a class of problems (those with an underlying generating function, such as a diffusion process). Note that for a network of actuated sensors the issue of sensor placement is in fact an issue of multi-agent coordination. Of course, such an issue is task-dependent. However, it is possible to define a general approach for coordination of multiple actuated sensors that are collected into a network in such a way that the approach can be applied to a variety of applications.

3. CONTROL-THEORETIC PROBLEM FORMULATION

In this section we formulate the general problem of coordination of distributed networks of actuated sensors for real-time spatial diffusion characterization. There is quite a bit of existing related work on coordination strategies for swarm-type networks, most of it based on the notion of individual sensors following some type of a-priori energy function or gradient. Here we propose a new idea: that of a model-based coordination strategy. One argument for this approach is that: a) models are becoming increasingly well-developed in a variety of application arenas, and b) one should use all the information that is available when trying to solve a problem. To describe our ideas about a network of sensors that can be (robotically) moved from one location to another we introduce the following terminology:

Actuated sensor: An actuated sensor S^A is a sensor that can move in space, either through self-logic or in response to a command from a supervisor.

Actuated sensor network: An (actuated sensor) network NS^A is a networked collection of actuated sensors, which are working together to achieve some type of information collection and processing.

Sensor Net: We begin by assuming we are given a network of actuated sensors, NS^A, made up of a collection of individual sensors that are defined as follows:

$S_i^A(q_i)$: an actuated sensor with the following characteristics:

- located in space at $q_i(t) = (x_i, y_i, z_i)^T \in R^3$
- can communicate with all others and with a supervisor.
- can generate a measurement of interest to the application, defined by $s_i(q_i, t)$, which is assumed to be a function of both space and time.
- can move freely in three dimensions with dynamics given by $\dot{q}_i = f_i(q_i, u_i)$, where $u_i(t)$ is the motion control input for sensor $S_i^A(q)$.

System to be Characterized: Next, we assume that there exists a space-time distribution of interest that we wish to characterize with the distributed actuated sensor network. We denote the distribution as $V(q, t)$, which is assumed to be the solution a known PDE with a known initial condition $V(q_0, t_0)$. The plant dynamics are assumed to be of the following form (which takes into account diffusion and transport phenomena effects such as convention/advection), expressed in standard vector calculus:

$$\frac{\partial V(q,t)}{\partial t} + \Delta \cdot (FV(q,t)) = \Delta \cdot (D(q,t)\Delta V(q,t)) + g(q,t)$$

$$V(q_0, t_0) = V_0$$

where $FV(q,t)$ denotes the effect of external, possibly variable, "inputs" on the plant dynamics (e.g., wind, rain, dust, humidity, etc.), $D(q,t)$ is the diffusion function for the specific problem, $g(q,t)$ reflects the effects of constraints (e.g., gravity, buildings, terrain, etc.), and V_0, q_0, t_0 denote the initial conditions.

Sampling Action: It is assumed that the output of the sensor $S_i^A(q)$, defined above as s_i, is a measurement of the distribution of interest at wherever the sensor is located in space. Thus we can write:

$$s_i(q_i, t) = V(q_i, t)$$

Prediction: The next step in the problem formulation is to define the prediction. Of course, if we had perfect knowledge, the problem would be trivial. However, in fact we only have estimates of the initial conditions, of the external inputs, and of the constraints. Let is define these estimates as $\hat{F}\hat{V}(q,t)$, $\hat{g}(q,t)$, and $\hat{V}_0$, respectively (of course, there are other sources of uncertainty, such as parameters in the diffusion function $D(q,t)$, but for now we will assume these are known). Then we can compute the estimated diffusion $\hat{V}(q,t)$ as the solution of

$$\frac{\partial \hat{V}(q,t)}{\partial t} + \Delta \cdot (\hat{F}\hat{V}(q,t)) = \Delta \cdot (D(q,t)\Delta\hat{V}(q,t)) + \hat{g}(q,t)$$

$$\hat{V}(q_0, t_0) = \hat{V}_0$$

$$\hat{V}(q_i, t_{s_i}) = s_i(q_i, t_s) = V(q_i, t_{s_i}) \text{ for all } i \text{ and all } t_{s_i}$$

Notice the introduction of the actual sensor measurements at sample points and sample times (q_i, t_{s_i}) as constraints for the partial differential equation.

Control: The next piece we add in this section is the motion control of the actuated sensor. There are various ways to approach this piece. For instance, one could take control actions to be a function of the error between the predicted samples and the actual samples. That is, given a set of samples, we make a prediction about the distribution. We then move to a new point in space and take new samples. The error between what we expect to measure and what we actually measure determines where we take our next samples. However, for the moment we consider a simpler approach. We simply move the sensors so they are uniformly distributed relative to the predicted distribution. Thus, we can write:

$$q_i^{sp} = H(\hat{V}(q,t))$$

$$\dot{u}_i = h_i(q_i^{sp} - q_i)$$

where H denotes a type of feed-forward energy function that has the effect of computing a set of uniformly distributed locations around the distribution and h is the control law used to drive the actuated senor to its new setpoint q_i^{sp}.

Goal Statement: Finally, we need to define the goal of the control action. Ideally, one would like to achieve

$$\lim_{t \to \infty} \hat{V}(q,t) = V(q,t) \text{ for all } q$$

However, this is quite ambitious. Instead, it may be better to hope for making the prediction match at the sample points. Thus, we can define a cost function

$$J = \lim_{t \to \infty} \sum_i h_g((\hat{V}(q_i,t) - V(q_i,t)))$$

where $h_g(\cdot)$ is a positive function

We comment that the design freedom in the problem is found in the selection of the controller motion functions H and h_i. For the most part, the selection of h_i is a straightforward control system design activity that will be specific to the robotic strategy used to actuate the sensor. The selection of H is ultimately the major design effort in the problem.
Finally, we note that although this is essentially an open-loop system identification problem, there is in fact a feedback feature to the problem, due to the motion control coupling to the output of the predictor.

4. DISTRIBUTED NETWORKS OF MOBILE (ACTUATED) ACTUATORS AND SENSORS

Next we propose the concept of mobile actuators as well as mobile sensors. Fig. 4 illustrates the concept in the context of the motivating application described in Section 2.1. In this figure, we show mobile, or actuated, actuators being deployed to release a dispersal agent into the contaminant plume. Of course, such actuators might be co-located with the actuated sensors. But, in many applications dispersal or actuating agents will typically be much more expensive than sensing agents. Thus it is reasonable to consider them separately. The effect of such actuated actuators can be added to our problem by noting that the primary effect they introduce is that of modifying the diffusion function in the system dynamics. Then we proceed as follows:

Actuator Network: We begin by assuming we are given a network of autonomous, actuated actuators defined as follows:

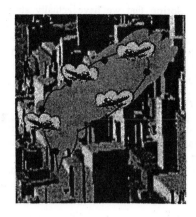

Fig. 4. Actuated actuator network.

$A_j^A(q_j)$: an actuated actuator with the following characteristics

- located in space at $q_j(t) = (x_j, y_j, z_j)^T \in R^3$
- can communicate with all others, with all sensors, and with a supervisor.
- can generate an effect of interest to the application, defined by $d_j^a(q_j, t)$, which is assumed to be a function of both space and time.
- can move freely in three dimensions with dynamics given by

$$\dot{q}_j = f_j^a(q_j, u_j^a)$$

where $u_j^a(t)$ is the motion control input for actuator j.

For this set of actuators we define a motion controller given by

$$q_j^{sp} = H^a(V^d(q,t) - \hat{V}(q,t))$$

$$\dot{u}_j^a = h_j^a(q_j^{sp} - q_j)$$

We point out that in the case of an actuated actuator, the function H^a is primarily a comparator and $V^d(q,t)$, the desired distribution, can typically be taken as zero (i.e., we don't want any contaminant!).

Control Goal: Without going into the details, we propose the following cost function for the design of the functions H^a and h_j^a

$$J^c = \lim_{t \to \infty} \int (V^d(q,t) - \hat{V}(q,t)) dq$$

This costs seeks to drive the predicted distribution to the final distribution everywhere in space.

Final Architecture and Problem Statement: The equations shown in Box 1 give the final form of the problem. Notice that we have actually stated two coupled problems. The sensor motion control problem is based on the output of the prediction. But the effect of the actuated actuators is shown in the diffusion function used in the prediction. We denote this as function

$$w(D(d,t),d_j^a(q_j,t))$$

because in general the effect of a dispersal agent may not necessarily be linear. At this time the effect of this coupling is not clear. One would hope to see the standard separation principle emerge, but that may not be possible. Deep research is needed to understand this problem.

CONCLUSION

In this paper we have presented new ideas related to the interpretation and use of data in distributed sensor networks, with a particular focus on coordination strategies for networks of actuated sensors. We first introduced some of the research issues related to such networks. We then described our idea (of using a model-based approach to sensor coordination) in the context of a specific application scenario: the real-time mapping of contaminant plume development. Next we formulated the problem in a general control-theoretic framework, which we also extended to include the notion of a network of mobile actuators combined with a network of mobile sensors. Current work is aimed at considering a variety of theoretical problems associated with the complete problem statement of Box 1 as well as at developing an experimental testbed using mobile robots and a simple fluid-based diffusion process.

REFERENCES

Abdelzaher, T., Stankovic, J., Son, S.; Blum, B., He, T., Wood, A., and Lu, C., (2003). A communication architecture and programming abstractions for real-time embedded sensor networks. *Proceedings of The 23rd International Conference on Distributed Computing Systems Workshops,* 2003. pp. 220-225.

Akyildiz, I.F., Weilian Su, Sankarasubramaniam, Y., and Cayirci, E, (2002). A survey on sensor networks. *IEEE Communications Magazine,* vol. 40, no. 8, August 2002, pp. 102-114.

Cerpa, A., J. Elson, D. Estrin, L. Girod, M. Hamilton and J. Zhao, (2001). Habitat monitoring: Application driver for wireless communications technology. *In 2001 ACM SIGCOMM Workshop on Data Communications in Latin America and the Caribbean,* San Jose, Costa Rica, Apr. 2001.

Haenggi, M. (2002). Mobile sensor-actuator networks: opportunities and challenges. *Proceedings of the 7th IEEE International Workshop on Cellular Neural Networks and Their Applications,* 2002. (CNNA 2002), 22-24 July 2002, pp. 283 -290.

Hairong Qi, S. Sitharama Iyengarb and Krishnendu Chakrabarty, (2001). Distributed sensor networks- a review of recent research. *Journal of the Franklin Institute,* vol. 338, , pp. 655–668.

Sukhatme G.S., Estrin D., Caron D., Mataric M. and Requicha A. (2000). Proposed Approach for Combining Distributed Sensing, Robotic Sampling, and Offline Analysis for in situ Marine Monitoring, in *Proc. Advanced Environmental and Chemical Sensing Technology - SPIE 2000,* Vol. 4205 Boston, November 6-8 2000.

Vladimir Bychkovskiy, Seapahn Megerian, Deborah Estrin, and Miodrag Potkonjak (2003). A Collaborative Approach to In-Place Sensor Calibration. *Proceedings of the 2nd International Workshop on Information Processing in Sensor Networks* (IPSN'03), also in *volume 2634 of Lecture Notes in Computer Science,* pp. 301-316, Springer-Verlag, 2003.

Wang, H., Elson, J., Girod, L., Estrin, D., and Yao, K. (2003). Target classification and localization in habitat monitoring. *Proceedings of the 2003 IEEE International Conference on Acoustics, Speech, and Signal Processing.* (ICASSP '03). Vol. 4, April 6-10, 2003, pp. IV_844-IV_847.

Whitehouse, K. and D. Culler (2002). Calibration as parameter estimation in sensor networks. *Proceedings of the 2002 ACM International Workshop on Wireless Sensor Networks and Applications,* September 28, 2002, Atlanta, GA.

Yang, H. and Sikdar, B, (2003). A protocol for tracking mobile targets using sensor networks. *Proceedings of the First IEEE International Workshop on Sensor Network Protocols and Applications,* 2003, pp. 71-81.

BOX 1

$$\min_{H,h} J^p = \lim_{t \to \infty} \sum_i (V(q_i,t) - \hat{V}(q_i,t))$$

$$\min_{H^a,h^a} J^c = \lim_{t \to \infty} \int (V^d(q,t) - \hat{V}(q,t))dq$$

subject to :

1a) $\quad \dfrac{\partial V(q,t)}{\partial t} + \Delta \cdot (FV(q,t)) = \Delta \cdot (D(q,t)\Delta V(q,t)) + g(q,t)$

1b) $\quad V(q_0,t_0) = V_0$

2a) $\quad \dfrac{\partial \hat{V}(q,t)}{\partial t} + \Delta \cdot (\hat{F}\hat{V}(q,t)) = \Delta \cdot (w(D(q,t),d_j^j(q_j,t))\Delta \hat{V}(q,t)) + \hat{g}(q,t)$

2b) $\quad \hat{V}(q_0,t_0) = \hat{V}_0$

2c) $\quad \hat{V}(q_i,t_{s_i}) = s_i(q_i,t_{s_i}) = V(q_i,t_{s_i})$ for all i and all t_{s_i}

3a) $\quad \dot{q}_i = f_i(q_i,u_i)$

3b) $\quad s_i(q_i,t_{s_i}) = V(q_i,t_{s_i})$

4a) $\quad q_i^{sp} = H(\hat{V}(q,t))$

4b) $\quad \dot{u}_i = h_i(q_i^{sp} - q_i)$

5a) $\quad \dot{q}_j = f_j^a(q_j,u_j^a)$

5b) $\quad q_j^{sp} = H^a(V^d(q,t) - \hat{V}(q,t))$

5c) $\quad \dot{u}_j^a = h_j^a(q_j^{sp} - q_j)$

Copyright © IFAC Telematics Applications in Automation and Robotics, Espoo, Finland, 2004

Multientity Rescue System

Jari Saarinen*, Jussi Suomela*, Aarne Halme* and Jirí Pavlícek**

* Helsinki University of Technology, Automation Technology Laboratory,
PL5400, 02015 HUT, Finland, Email: jari.saarinen@hut.fi
** Czech Technical University, Gerstner Laboratory,Technicka 2, CZ 166 27, Prague 6, Czech Republic

Abstract: This paper presents methods for a cooperative rescue operation in partially or totally unknown areas with human and robotic explorers. Cooperation is based on a common presence i.e. a common understanding of the environment for both humans and robots. The common presence is generated by continuous mapping data supplied by robotic and human entities. Mapping data is pre-processed and transmitted to the mission controller/operator, who finally filters the data to a common presence, which is transmitted continuously back to both entities. The studied methods are human navigation without artificial beacons, human and robotic SLAM, cooperative localization and cooperative map/model building for common presence. Methods are developed, tested and integrated in a European Community research project called PeLoTe. *Copyright © 2004 IFAC*

Keywords: Augmented reality, telematics, human navigation, telepresence, localisation, mapping, SLAM

1. INTRODUCTION

The main difficulty in the human robot cooperation and the interface design is the different level of cognition. In order to work and communicate well together, robots and humans should understand the same language, its contents and abstractions and the common environment similarly. The processes of the human cognition – especially the understanding level – are studied a lot but they are still practically unknown. On the other hand we know very well what our cognition can do. Both functions and capability of robot/computer cognition is known well because we have created them. As a result we can be sure that both the function and performance of human and robot cognition are totally different.

In rescue situation the key issue is the environment. When robots and humans are working in same area they should be able to understand the common environment – have *a common presence* - and change information relative to it. Especially important this is in mapping type of tasks where both robots and humans generate information of the environment and both should be capable to exploit the information generated by the other. The telematics offers the possibility to fuse the information and create continuously updated presence model.

1.1 Rescue scenario

Pelote (Building Presence through Localization for Hybrid Telematic Systems) is part of the IST programme of the European Community [1]. The target of the project is to study how to map a totally or partially unknown area with a group of human and robotic entities and form a common environment model (presence) from the mapped data produced by both entities. The model will be updated in real-time and it provides presence for both humans and robots. This type of scenario is typical in rescue, military and planetary exploration tasks.

The case example in Pelote is a rescue task where firemen and supporting robots are mapping a common area together with the help of a remote operator Fig. 1. Both entities specialize to tasks, which are natural for them. Robots can perform accurate navigation and measurements from the environment even in hostile conditions. Humans can give fast, informative verbal descriptions of the situation and conditions. Human senses are also more versatile than robot senses and the human cognition is superior in perceiving and describing entireties to another human. Exploring entities are supported by a remote human operator in a mobile control room. He teleoperates the robots and supervises the firemen. He also summarizes the information obtained from both entities to the common environment model. Dataflow between the operator and entities is illustrated in Fig. 2.

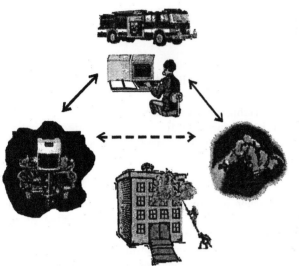

Fig. 1. Pelote scenario: Human and robotic entities are exploring and mapping a common area.

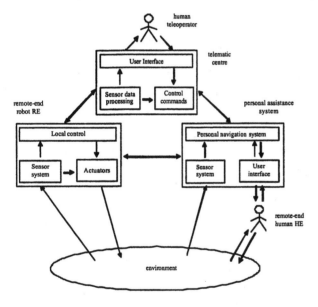

Fig. 2. Dataflow and communication in Pelote scenario

2. COMMON PRESENCE

The main topic of the project is to generate a presence from the data perceived by human and robotic entities. The presence should be common for the both entities. The common presence consists of 2D or 3D map or model and an object database, which can be updated with any kind of object information. The model is composed from possible apriori map.

The model shows the navigable area and borders like walls, constant objects etc. All additional information and mapped objects are stored in the database, where they can be picked up and augmented in the model. The continuous mapping data from robotic and human explorers is added to the database. The model is shown as a topographic map to humans and as a 2D occupancy grid to robots.

The operator can afterwards include or correct "constant wall – type" objects as part of the model. In the database all objects have at least size, position, type, additional information and layer group. Additional information can be anything, even a photo or verbal description, which can be seen/hear by clicking the object. The layer group information is used to visualise only the needed information in order to avoid the information flood on the model.

2.1 Description of common presence and virtual working environment

In PeLoTe project the goal is to prove that through localization we can build a common presence. However, it must be underlined that the presence is for Hybrid Telematic entities. The different entities have different understanding of the world. The hypothesis of the project is that the *common presence* for these entities is the location (spatial) awareness.

The term *common presence* means that all entities have some common space, which they can understand in similar way and exchange information through it. The common presence can be understood as a virtual working environment for different types of entities. The objects in the virtual environment are understandable to all entities. This kind of virtual space does not satisfy all the components of the virtual environment for a human, but there are some important key features that are satisfied. Especially the key phrase "Being there" is satisfied. All entities have a location, which puts them inside this virtual space. All entities have the capability to modify the environment, through mapping, inserting new objects etc.

Also through the virtual environment the social presence is somehow fulfilled. All the entities can be seen in the environment, they are there. Further more, different entities can communicate through the virtual environment (e.g. a robot can notify that it has found a victim, and a human confirms it through robot sensors or the operator can guide the human by displaying the position in the screen).

To create this kind of common working environment for humans and robots, there must exist some common base understanding between these entities. This understanding is foreseen to be the *spatial awareness*. The term spatial awareness means that we can understand the environment in the same way. One essential feature of the spatial awareness is to share own position inside the virtual environment. The accurate position inside the virtual world gives means to update (modify) this world. For example, the robots can build this world, by using environmental sensors. So in the end the common presence is a spatial description of the working environment, with added common objects that can be understood by all entities.

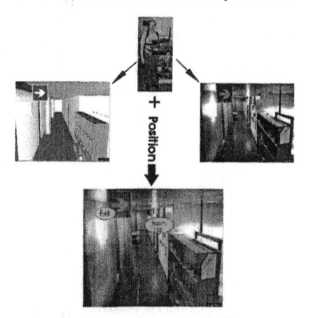

Fig 3. Creating a virtual working environment for a human. The position information makes it possible to fuse information from the model and real environment. The outcome is fused as augmented reality, where the extra information about the environment can be added [3].

This common presence can be seen as one step forward in the user interface design. The key features of the presence are that a human can for example control the robot by using high-level commands. This requires that both the robot and the human understand the environment similarly.

The PeLoTe demonstration scenario is a rescue scenario, which includes one operator (remote), one human and two robots inside the building. The situation now, without our system would be following: the operator is outside the building and has only a voice link between the human teams inside. The operator and the firemen might have a rough map of the building they are operating in. In the daylight, with no smoke an experienced team of fire fighters with predefined mission and skilful operator control will operate very well without problems. But when the building is completely dark or full of smoke the degree of difficulty increases. The fire fighters cannot use their visual sense anymore and the self-localisation gets much harder – in practise it will be impossible. This affects also

to the performance of the mission, it is no longer clear where the different people have been, and where they should go. Also the operator has to trust only what he hears from the fire fighters and try to give further advices. Since nobody has a clear idea what is happening, the mission relies more or less in the imagination and experience of the fire fighters. The operator certainly cannot say that he is present in the building by having only audio information. One interesting point to note is that also the firemen inside the building have different levels of presence. These feelings of presence are something completely different from operator's one. The firemen are actually working in the physical environment, but with limited senses (e.g. no visual input due to the darkness).

With PeLoTe "virtual working environment" system, everybody is located inside the virtual world, which describes the real world. They can see all the time where the people move and where they have been. For operator this gives valuable information. He can direct the people in the scene if needed, and everyone can be pinpointed. Also for people on the scene this is important, now they see where they are situated in the map and can use this information to move to the next target i.e. they can navigate.

2.2 Standard Rescue Map (SRM)

The apriori information from a rescue area is called SRM. The future objective is that SRMs of all buildings with straight connection to the public alarm centre will be in the database of the rescue officials. After an alarm the firemen could already during their transfer to rescue place make planning with the help of standard form maps.

SRM is a 2D or 3D (if feasible) model and an object database. It creates also the base for the common presence as described in the previous chapter. The model shows the navigable area and all important objects - from the rescue point of view - like sprinklers, fire alarm areas, flammable or poisonous chemicals etc. are stored in the database. A good example of similar type of approach is the ECDIS map system used in ships [2]. However, ECDIS is limited in 2DoF.

3. HUMAN ENTITIES (HE)

3.1 Personal Navigation (PeNa)

In a rescue situation it's essential to know the position of all entities. Unlike robots humans only rarely know their accurate position. Only in situations when human can identify a known object, like corridor crossing or stairway, he can know his accurate position relative to that object. Personal navigation (PeNa) under development is intended for continuous automatic localization of a human. The main challenge in the Pena development is the functionality without any existing beacons like GPS, WLAN and cellular base stations. Pena is based on human dead reckoning supported by radio beacons (if available), and the human's own senses and perception. Also the robots can support Pena. Dead reckoning includes step measurements (pedometer), a lab made stride length measurement unit (Silmu), magnetic sensors (compass) and inertial measurements (gyro and accelerometers). Robots have relatively better position information of their position than humans. To help the human positioning robots can either carry or drop the wireless (radio/US) beacons, which humans can use to improve the positioning accuracy. Additionally, the feasibility of human SLAM based on laser scanner is studied.

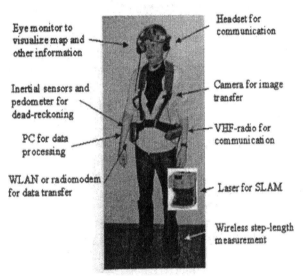

Fig. 4. Human hardware, the "personal assistance system" including PeNa

In addition to PeNa hardware, the human entities are equipped with data/speech communication links and a HMD, which provides the real-time mapping information with own position. This whole system is called Personal Assistance System (PAS) illustrated in fig. 4.

3.1.2 Cooperative localization

Despite of how well the human dead reckoning will succeed, there will always be an accumulative error, which can grow very rapidly in special situations like during creeping or climbing over an obstacle. The only way to correct this accumulative error is to use beacons to correct the error always when possible. In the case of an apriori map from the environment, human can use visual beacons

to correct his position. HE sees his position on the map from the HMD. If the error is noticeable he can identify a known object, like corridor crossing or stairway, and simply describe his real position and ask the operator to correct the position. However, in most cases humans don't have time to do this or the map is not available. When external beacon systems like GPS, cellular or WLAN base stations are not available indoors, a mobile beacon system based on robots is under developed.

The cooperative localization is based on the fact that robots know their position much better than humans. Robots have more accurate dead reckoning comparing to humans and they will also perform continuous SLAM, which produces excellent position information especially in the case of an apriori map. Robots will be equipped with radio/US beacons (fig 5.), which can either be onboard all the time or dropped to important places like corridor crossings.

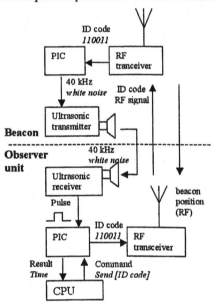

Fig. 5. The principle of cooperative localization based on radio/US beacons

3.2 Operator

The tasks of the remote teleoperator are the integration of the information provided by different teams, the coordination of the remote team members with respect to global objectives, provision of background information from data bases to support the remote human team members to perform their task and supervisory control of the remote robot team members. In order to accomplish these tasks the teleoperator needs a sophisticated, intelligent and informative user interface.

The user interface is based on the environment model/map. The base of the UI is the a-priori knowledge, standard rescue map SRM, that is based on the known information of the building. The information is classified to different layers, so that irrelevant information can be fused out in different situations. The other part of the UI is collected from the mission data. This part is called dynamical part and is also built layered manner (e.g. each entity has an own layer that consists of entity position, history path and planned path) (see fig. 6).

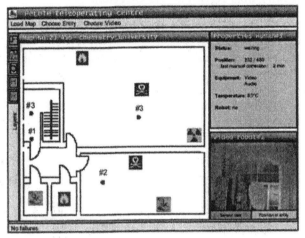

Figure 6. User interface design

3.3. Robot entities (RE)

The robots (Fig. 7) are used for accurate mapping and perception of the unknown/destructed areas. They will also explore dangerous areas. The robots are teleoperated by the operator, who points the zones to be explored or even takes a robot under direct control. Robots do simultaneous localization and mapping (SLAM) based on the odometry, a laser scanner and other possible sensors. Robots also utilize the cooperative localization beacons. The pre-processed perception data and camera images are transferred continuously to the operator and the updated map data is received from the operator.

Fig. 7. Robots Merlin and J2B2

4. COMMUNICATIONS AND DATA SHARING

4.1. HE – Operator interface

The position data from Pena and the mapping data from the operator are transferred automatically via a wireless interface. The actual communication between the human entities and the operator is mainly done by speech. It's assumed that especially in rescue operations the men don't have time to play with computer interfaces but they can speak simultaneously as they work. For example when a fireman notices that a remarkable error has accumulated in his position on the HMD, instead of correcting his position by using a trackball, he asks the operator: "I am front of the elevator facing to main door, please correct my position". Now the operator makes the correction and the fireman can check the updated position in real time from his display. This is of course possible only when an a-priori map is available. When robots are using laser scanners and other sensors for accurate numerical mapping, humans use their ability to piece together wholenesses. Again a fireman acts as an intelligent sensor and asks the operator to update the map. For example "the corridor in front of me is collapsed" and the operator marks the corridor as a non-enter area in the map.

4.2 RE – operator communication

Robots work under supervision of the operator. Control is based both on task based (supervisory) control and direct teleoperation, which demands a short delay communication. Additionally to the control robots send pre-processed mapping data to the operator and receive the updated environment model (presence).

The mapping and presence information is transmitted to/from the robots as occupancy grid maps, which are readable for both the humans and the robots. However, for the HEs the presence model is transmitted as geometric maps always when it's possible. To support the presence of the operator the robots provide also video and audio data from the rescue area.

4.3 HE – RE communication

The communication between the rescue entities goes mainly via the operator. As an optional feature the flexible communication hardware provides a possibility for the straight communication also. Possible tasks could be a direct robot teleoperation controlled by one of the human entities or a so-called dog-robot, which is continuously tracking its master, taking direct commands and providing sensor (and position) data to the master.

4.4 Communication Architecture

The goal of the project requires sharing out the knowledge between the hybrid telematic entities. The problem in rescue situations is that a ready-made infrastructure cannot be expected. The problem of the communication media is, however, not in the scope of the project. The aim is to develop methods for robust data exchange so that the system won't fail even in the case of communication black outs. In the project WLAN is used as communication media. It provides enough bandwidth and an easy interface for prototyping.

There are basically three different types of modules interacting and changing information: the teleoperation centre (TC), the Personal assistance system (PAS) and the robot system. Figure 8 explains the interactions among the modules. TC is on the top of the system hierarchy and it's used by the human operator for controlling and monitoring the other modules. As can be seen from the Figure 8, there can be more instances of the TC modules as well as the robot and the PAS modules. One TC module can control multiple robot and PAS modules. If the number of the robot and the PAS modules is too high, then multiple TCs can be used to control separate groups of robot and PAS modules. In an extreme situation each PAS or robot module can be controlled by its personal TC.

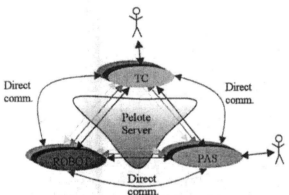

Figure 8: Interaction among PeLoTe module.

Due to the requirement of sharing the knowledge among all entities, some mechanism providing such a function is needed. The Client/Server architecture seems to be the most feasible way of sharing information among the entities. The server responds only to requests from the clients and it will never send an update to a client without a previous request from the client. Thus, the clients need to ask for the respective updates. In such architecture, the server does not have to take care about the communication; if it is broken the client has to re-establish it. When a client

need some periodically update, it should pool the server. Figure 8 also shows the domain of the PeLoTe Server. The shared information is stored in the PeLoTe Server and provided to the client applications on by request using Server Services. TC, Robot and PAS modules are considered as client applications. In the case of a communication failure with the server or in case of a fast response requirement, direct communication between modules can also be established (eg. direct control of robotic entity by human using PAS system or direct drawing from TC GUI to PAS GUI etc.).

The communication infrastructure in PeLoTe is based on Java RMI (Remote Method Invocation). Remote Method Invocation (RMI) technology, first introduced in JDK 1.1, elevates the network programming to a higher plane. Although RMI is relatively easy to use, it is a remarkably powerful technology. The RMI requires Java and a TCP/IP network. The WLAN network provides the TCP/IP automatically.

6. CONCLUSIONS AND DISCUSSION

While this paper is written the Pelote project is in the midway and part of the described methods like the human dead reckoning, the human SLAM and building augmented reality have been already tested and found feasible for the cooperative mapping and presence building task. Especially the laser scanner based human SLAM will be studied further as well as human dead reckoning will be improved. System integration, testing and final demonstrating will be carried out before the April 2005.

ACKNOWLEDGEMENTS

The work of all authors has been supported within the IST-2001-FET framework under project no. 38873 "PeLoTe". The support is gratefully acknowledged.

REFERENCES

[1] Suomela J, Saarinen J, Halme A, Harmo P, Proceedings of Online Interactive Building of Presence, The 4th International Conference on Field And Service Robotics, July 14-16, 2003.
[2] http://www.openecdis.org/
[3] Harmo P., Halme A., Pitkänen H., Virekoski P., Halinen M., Suomela J., Moving Eye – Interactive Telepresence Over Internet With A Ball Shaped mobile Robot, Proceedings of IFAC Conference on Telematics Applications in Automation and Robotics, TA2001, 24-26 July 2001, Weingarten, Germany

Copyright © IFAC Telematics Applications in Automation
and Robotics, Espoo, Finland, 2004

ELSEVIER

IFAC
PUBLICATIONS
www.elsevier.com/locate/ifac

MANAGING THE COMMUNICATION IN A COMPLEX SYSTEM OF ROBOTS AND SENSORS

Piotr Skrzypczyński

Poznań University of Technology,
Institute of Control and Information Engineering
ul. Piotrowo 3A, PL-60-965 Poznań, Poland

Abstract: This article discusses the problems of communication and information
sharing in a system of mobile robots and stationary monitoring sensors. A negotiation
framework based on the Contract Net approach supports communication between the
distributed sensing processes. The hybrid blackboard/message based architecture has
been defined to manage the communication within the agents and at the inter-agent
level. *Copyright© 2004 IFAC*

Keywords: Mobile robots, Distributed sensing, Wireless LAN, Software architectures

1. INTRODUCTION

In systems performing indoor transportation, service or patrol tasks several mobile robots cooperate with stationary monitoring cameras and other sensors (e.g. anti-intrusion, fire-detection), which are already integrated within a communication network. Information from the stationary sensors available in the environment can help the robots – these sensors can serve as external navigation aids sharing with the robots information about the state of the environment and yielding the pose of the vehicles. Such an information can significantly improve the navigation capabilities of the robots (Bączyk and Skrzypczyński, 2003).

Integration of stationary sensors and mobile robots within a distributed and dynamic system raises some specific issues as to the communication framework and the software architecture:

- efficient use of the communication channel(s);
- fusion of data from sensors with various abilities to extract features;
- consistent management of the transmitted information uncertainty;
- robustness to component failures;
- software complexity management.

In this article a system of heterogenous (with regard to perception abilities) mobile robots supported by stationary cameras is considered as an example application. The number and the properties of components may vary, thus the system must be open to modifications. The concept of agents is used to model mobile robots, monitoring sensors (overhead cameras), and the human operator interface. An agent has perception and communication abilities, and its functionality is expressed through the actions it takes, including the communication actions. The robot agents (**RA**) gather information from their own sensors, from other robots, from stationary devices, and construct the internal world model. Each robot uses its odometry and the on-board sensors to determine its own pose (position and orientation). If the pose uncertainty exceeds the acceptable value, the robot asks for positioning service from the external agents. Perception agents (**PA**) compete for serving positioning data to robots. They are based on the overhead cameras as the hardware components. Operator agent (**OA**) initializes, configures and monitors robots and perception agents.

Logical framework of the Contract Net Protocol (Smith, 1980) has been used to manage the negotiations between particular agents, which need the navigation-related information or which offer such data to the others. Particular modules (programs) in the system communicate via an IP-based LAN. A dedicated communication system has been implemented to run over mixed wired/wireless LAN (Kasiński and Skrzypczyński, 2002).

2. INTRA-AGENT COMMUNICATION

In the mobile robots and the monitoring sensors a multi-stage and multi-source data processing is undertaken. Each of the data processing operations can be separately defined as a black box with some input and output. These black boxes are loosely coupled by data they exchange. This kind of data processing can be organized as a blackboard system with a shared database and a set of experts cooperating in a data-driven and opportunistic way. In turn, the blackboard system can be modeled and implemented as an agent system with experts working as agents.

The software running on the robot-agents is based on the multi-agent blackboard architecture introduced in (Brzykcy et al., 2001). The sensor/actuator drivers, data fusion, and data transformation modules work as software agents communicating through the blackboard. The blackboard contains different descriptions of the robot environment and task. Blackboard agents are coupled to physical devices – sensors and actuators or to processing tasks – experts. The device agents execute their actions concurrently, observing time constraints of respective sensors and actuators. Around the blackboard, there are device agents representing sonars, the laser scanner, the onboard camera, and the robot controller. The data processing blackboard agents are: the map building agents, the self-localization agent, and the pilot agent providing the behaviour-based layer of the navigation system. Monitoring of the whole dataflow and the execution of operator commands is the duty of the report agent. Blackboard agents detect events in the system by observing the changes of data on the blackboard. The information needed to arrange control is implemented by means of specialized flags. The blackboard agents cannot communicate directly each-other or with other robots and perception agents. The internal communication within the robot's body relies on the blackboard, while the system-level, inter-agent communication is implemented by the specialized device agent, which is manager of the physical communication channel (LAN).

The perception agents localize the robots with respect to the global reference frame. The robots are equipped with LED markers to simplify the positioning task and to improve the reliability of the robot detection under varying illumination conditions (Bączyk and Skrzypczyński, 2003).

The software architecture of the perception agent is based on the same considerations regarding the data-driven processing of information as the architecture of the robots. The blackboard holds all the data, which are accessed by the camera/frame-grabber device agent, the image processing and robot positioning agents, and the manager of the communication channel. Monitoring and command execution is performed by the report agent.

Due to the unification of the internal architectures all agents in the system can share components, such as the communication channel manager, also the substitution of physical components of the agent, e.g. the frame-grabber type is relatively easy.

3. INTER-AGENT COMMUNICATION

Although the architecture of the agents working in the system is blackboard-based, a message-passing approach to the inter-agent communication has been adopted to avoid problems with the management of a global blackboard. A single blackboard existing in a system distributed over physically separated nodes can easily become a bottleneck.

The agents understand the messages using the predefined list of message types, which is known to all of them. The message is a tuple: Msg=(ID, REC, SEN, TYP, LEN, DAT), with the following meaning of fields: ID – unique message number; REC – receiver address (agent ID); SEN – sender address (agent ID); TYP – message type; LEN – message length; DAT – task-specific data (string of bytes, may not exist in some types of messages). The types of messages correspond with the interpretation and meaning of the DAT field.

The agents use a dedicated protocol based on the Contract Net concept (Smith, 1980) to choose the best data from the available external sources. The agent, which is interested in pursuing a task advertises the existence of this task to other agents by sending them a request message. Agents that are eligible to participate in the negotiation (bidders) evaluate their ability to perform the specific task, and send their bids for the contract to the agent, which has initialised the whole process, and now is the manager for it. The manager evaluates the bids from particular agents and awards contract for execution of the task to the agent with the highest declared ability to complete this task (e.g. the winner could be an agent, which offers the data with the lowest uncertainty). Manager and contractor are linked by a contract and they communicate in peer-to-peer mode to establish the transfer of information.

There are some tasks, which can not be executed by a single agent, because information from several agents is needed to complete the task. In such cases, the contractor may split up the task and

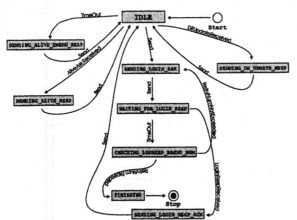

Fig. 1. Communication protocol finite state automaton – regular agent

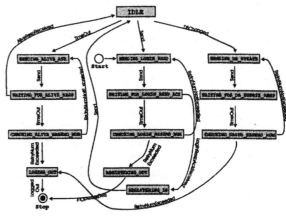

Fig. 2. Communication protocol finite state automaton – operator agent

award contracts to some other agents (such as an overhead camera-based perception agent). The agents operating in the presented system are not *a priori* designated as manager or bidder. Some of them can take both roles, depending on the current task context. Agents, which can not process data to yield results in the requested time (e.g. due to a sensor failure) are not considered as potential contractors by an agent that issued that particular task. For example, a robot-agent, which wants localization data from camera-based perception agents can award the contract to an other robot equipped with the on-board camera, whenever in the particular situation reliable pose estimate can not be provided by the overhead cameras.

The communication mechanism is based upon the User Datagram Protocol (UDP), a transport layer protocol simpler than TCP. UDP packets can be delivered to the appropriate process on the destination network node, and include a checksum ensuring correctness of the received data. Although TCP/IP communication is more reliable, this is achieved at the expense of a higher overhead and latency (Coulouris *et al.*, 1996). In a message-based communication system for robots, running in a single LAN, the characteristics of TCP – continuous data stream orientation and transparent recovery capabilities, being advantages in typical Internet applications, become rather disadvantages, especially when wireless LAN is being used (Gage, 1997).

At the start of the system, the address of the Operator Agent is known to all agents (stored in their configuration files). An agent entering the system contacts the **OA**, sending its symbolic name, type (**RA** or **PA**), IP address, and port number. This message includes also a short description of the agent's capabilities in the form of a list of specific tasks the agent can perform. The following task attributes have been specified: `ovr_loc_active` (for **PA**), `ovr_loc_passive`,

`cam_loc_active`, `cam_loc_passive`, `map_building` (for **RA**). The **OA** verifies the formal consistency of the data (e.g. an agent declaring itself as **PA** cannot have `ovr_loc_passive` on its list of tasks, because it cannot be observed by any other agent), and registers the new agent, giving it a unique ID. The operator updates its Agent Data Base (ADB) containing information on all agents, currently active in the system, and sends a copy of this ADB to all the agents. Hence, the knowledge of the current system configuration is distributed among all agents. Because the ADB contains also the specification of the tasks, which particular agents in the system can perform, it is possible to use point-to-point messages addressed to a subset of agents instead of broadcasting. Such an approach reduces the message traffic and allows to skip the eligibility specification in the CNP announcement messages. Figure 1 shows the above described communication protocol implemented in **RA** and **PA** as a finite state automaton.

In the case of a failure of the Operator Agent, a human supervisor can run another **OA**, which queries the necessary data from other agents. The task of the **OA** is also to periodically check if the agents are alive. It is accomplished by sending every 30 s the special message `M_ALIVE_ASK` to all agents registered in ADB. The agent who does not respond to this message several times, is considered "dead", and removed from ADB. The agents communicate by using point-to-point message passing. They use their local copies of the ADB to translate the symbolic names into real IP addresses. The data integrity is ensured by handshaking and timeouts. Figure 2 shows the finite state automaton of the protocol implemented in **OA**.

4. COMMUNICATION EFFICIENCY ISSUES

The software architecture and communication system reported here provide shared-memory-like communication for processes working locally (blackboard agents), and message-passing for communication between robots and perception agents.

With regard to the inter-agent communication, the aim was to develop a general procedure of "seeking advice" in such cases, when the robot cannot complete its task because of the lack of information (e.g. localization data) or particular skills (e.g. ability to recognize landmarks). The negotiation framework enables the robots to address the proper camera agent even if the robot does not know exactly where the external cameras are placed. This helps to make the distributed system open to modifications and robust to the failures of particular external sensors.

The possibility to share knowledge due to the explicit inter-agent communication facilitates certain tasks, but the advantages should be weighed against costs of the communication itself (Balch and Arkin, 1994), because the communication is an intentional act of the same class as sensing. The negotiation-based communication provides means to judge if it is worth to use the data transmitted from other agents instead of local data.

The chosen approach to communication is influenced by the hardware used in the system. The presented experimental system uses mixed wired/wireless Ethernet LAN of IEEE 802.3 and 802.11 standards. Over the last few years transmission rate and reliability in wireless networks have increased, also the 802.11-compatible hardware became cheaper and more popular in robotics applications. However, typical Ethernet network is not suitable for hard-real-time communication, mostly due to the non-deterministic handling of packets (Nett and Schemmer, 2003). This is because of the CSMA method used in the Media Access Control (MAC) layer, which allows collisions in the shared media. When such a collision occurs, both nodes wait some time before attempting to repeat the transmission.

The communication architecture proposed here avoids, to a large extend, the problems caused by the limitations of the underlying network. The agents which use network communication (**RA**, **PA**, **OA**) are loosely coupled, have high degree of autonomy, and do not rely on messages transmitted from/to other agents. Information gathered from cooperating agents is represented on the agent's blackboard in the same way as any sensor data or abstracted world model. The robot agents do not have modules/behaviors triggered by external messages, what is typical to many multi-robot systems heavily based on the behavioral paradigm. If an agent asks other agents for some data, and does not receive a response within a specified time amount it continues to act using the already available data (e.g. current pose estimate), and tries

to contact different agents. These properties of the communication scheme release the real-time constrains on the underlying network. The agents send messages in an opportunistic way, only when they need some external data. The messages are very short. For example, in the negotiation procedure for the case of the overhead camera-based positioning (Bączyk and Skrzypczyński, 2003), the request message has only 82 bytes, each bid from a potential contractor is 80 bytes long. This is possible because the blackboard-agents (data processing modules) have no individual communication capabilities, and are encapsulated within the **RA** and **PA** which communicate at relatively high level of data abstraction. Effectively, the messages consume only a small fraction of the network bandwidth, which is up to 11 Mbit/s for the IEEE 802.11b LAN.

The communication complexity (Fischer *et al.*, 1999) of the proposed negotiation procedure can be computed in a straightforward way. Let there are n agents which can become contractors for a given task. A manager sends n requests to these agents, each request being a single message. The agents response with n bids. Then, the contractor sends one acknowledgement and receives one message with the requested information. Hence, under an assumption that no agent failed to send the bid, there is $2n + 2$ messages, and the communication complexity is $O(n)$. The amount of data transferred depends on particular subject of negotiation, for example in the above mentioned localization task it is $82n + 80n + 18 + 82$ bytes. Adding the UDP, IP and MAC layer overheads, it makes $292n + 224$ bytes. Assuming 10 potential contractors (overhead cameras and robots with camera) the network transfers about 3 kB of data. The time needed to complete a negotiation session can be also estimated. Assuming small network load (no collisions) and typical time overheads imposed by the lower layers of the protocol stack (Calì *et al.*, 2000), the transmission times for the positioning can be computed as: transmission of a single request takes about 0.8 ms, transmission of a single bid is also about 0.8 ms, acknowledgement 0.5 ms, and 0.5 ms for the data transfer. Because the contractor sends requests as point-to-point messages the time to complete this negotiation phase for 10 agents is about 8 ms. In a worst case the contractor receives bids sequentially, for 10 bidders it takes also 8 ms. Hence, the whole transmission time needed to complete the negotiation for this example task is about 20 ms, what is quite small amount of time comparing it to the time needed for local sensory data processing in the agents.

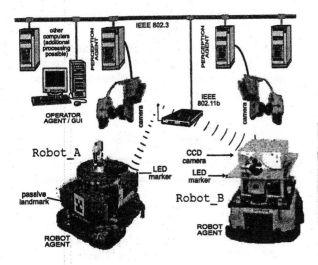

Fig. 3. Experimental set-up

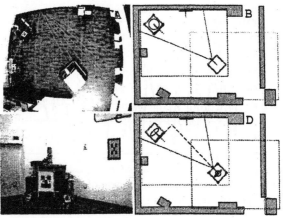

Fig. 4. Uncertainty management in the co-operative positioning

5. IMPLEMENTATION

In the experimental set-up, two mobile robots of Labmate type and two monitoring cameras attached to the ceiling are used (Fig. 3). Both robots have on-board PC computers, which are nodes of the LAN. One of the robots has a vision system with a fixed camera. Simple artificial landmarks with unique codes give the vision subsystem a chance to provide a reliable, alternative way of self-localization (Bączyk *et al.*, 2003). The robot with camera can also recognize the other robot (having a landmark), and can give it a "helping hand" by positioning it.

The following experiment demonstrates the ability of the system to fuse data from different sensors: stationary and hosted on the robots, to manage the transmitted pose uncertainty, and to recover from sensor failures. Figure 4A shows a situation in which Robot_A is located in a corner of the field of view of the perception agents, while Robot_B is near to the centre of this field. When Robot_A needs to know exactly its pose, it sends a positioning request to the agents, which according to the ADB content, are able to perform the loc_active tasks. The request contains the parameters $[x_0 \ y_0 \ \theta_0]^T$ of the current pose. One of the overhead-camera perception agents answers with a bid, but the predicted localization quality is low, because of the location of the robot (Fig. 4B). Another, potential contractor is the Robot_B, equipped with the on-board camera. It predicts the relative pose uncertainty (Bączyk and Skrzypczyński, 2003), and estimates the time needed to find and localize the landmark attached to Robot_A. However, to compute the pose of the robot with landmark in the global frame, the robot with camera needs to know its own pose. Although Robot_B can use the artificial landmark attached to the wall to compute this pose, the predicted positional uncertainty is quite high, and Robot_B sends a positioning request

to the perception agents. The Robot_B receives a bid from the perception agent containing the predicted positional uncertainty, and accepts it because this uncertainty is low due to the robot position under the overhead camera. The robot with camera computes the final predicted estimate of Robot_A pose, then sends the bid. The robot with landmark evaluates the received bids and accepts the one from Robot_B by sending the acknowledgement message. The contractor finalizes the contract with the perception agent receiving the current pose estimate with uncertainty information. Then, it performs the actual positioning procedure by taking the image, recognizing the landmark attached to the Robot_A (Fig. 4C), and estimating the pose of this robot with respect to its local frame. At the final step, Robot_B computes the pose estimate $[x_n \ y_n \ \theta_n]^T$ of Robot_A in the global frame (Fig. 4D).

To quantitatively demonstrate how the negotiations between the robot agents and the perception agents improve the positioning quality, results of two experiments have been compared. The robot followed a path relying for positioning only on its odometry and on the external camera agents. Figure 5 compares the positioning results as a function of the travelled distance for the two cases: with (solid line) and without (dashed line) the CNP-based negotiation procedure. The positional uncertainty ellipse area is used as the performance measure. When the robot did not use CNP, but asked the first available perception agent to localize it, the number of positioning requests was much bigger, and many of them were unsuccessful. In many cases the robot requested the positioning at the border of the field of view of the agent, where the position uncertainty is considerably higher (Bączyk and Skrzypczyński, 2003). The bid evaluation ensures that the robot uses the best positioning service offered within the system.

The time needed to complete the negotiations between a robot and the perception agents has

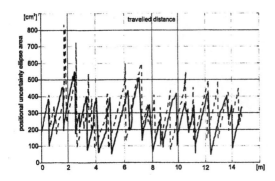

Fig. 5. Experimental comparison of the **PA**-based positioning results

been evaluated experimentally. Because the experimental set-up available in the lab has rather small scale, other stationary computers being nodes of the same LAN have been used to simulate additional perception agents involved in the negotiations. Results of this experiment are plotted in Fig 6, which shows average time needed to complete one round of the negotiations concerning the localization task as a function of the number of perception agents active in the system. The transmission times are greater than the one computed upon theoretical considerations, but a real network does not fullfil all the assumptions (collisions and lost packets are present), and the overhead of the local computations must be taken into account.

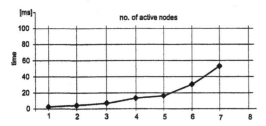

Fig. 6. Experimental evaluation of the transmission time

6. CONCLUSIONS

The communication is one of the most important aspects of the distributed robotic systems. This article describes an approach to fulfill the basic requirements imposed onto the communication framework in the multi-robot patrol/surveillance applications.

The main contribution of the described work is the new architecture for distributed multi-robot and multi-sensor systems, merging the advantages of the blackboard for the local communication with the message-passing backbone, which makes the whole system easily extendable and robust to failures. The efficiency of the proposed approach has been discussed extensively and shown on experimental results.

7. ACKNOWLEDGMENTS

The author benefited a lot from the discussions with K. M. Romanowski. Thanks go also to the students who contributed to implementation and testing of the system.

8. REFERENCES

Balch, T. and R. C. Arkin, (1994). Communication in Reactive Multiagent Robotic Systems, *Autonomous Robots*, 1, 1–25.

Bączyk, R., A. Kasiński and P. Skrzypczyński, (2003). Vision-Based Mobile Robot Localization with Simple Artificial Landmarks, *Prepr. 7th IFAC Symp. on Robot Control*, Wrocław, 217-222.

Bączyk, R. and P. Skrzypczyński, (2003). A Framework for Vision-Based Positioning in a Distributed Robotic System, *Proc. European Conf. on Mobile Robots*, Warsaw, 153-158.

Brzykcy, G., J. Martinek, A. Meissner and P. Skrzypczyński, (2001). Multi-Agent Blackboard Architecture for a Mobile Robot, *Proc. IEEE/RSJ Int. Conf. on Intelligent Robots and Systems*, Maui, 2369–2374.

Calì, F., M. Conti and E. Gregori, (2000). IEEE 802.11 Wireless LAN: Capacity Analysis and Protocol Enhancement, tech. rep., CNUCE Network Group.

Coulouris, G., J. Dollimore and T. Kindberg, (1996). *Distributed Systems. Concepts and Design*, Addison Wesley.

Fischer, K., B. Chaib-draa, J. Müller, M. Pischel and C. Gerber, (1999). A Simulation Approach Based on Negotiation and Cooperation Between Agents: A Case Study, *IEEE Trans. on Systems, Man, and Cybernetics*, 29(4), 531–545.

Gage, D. W. (1997). Network Protocols for Mobile Robot Systems, *SPIE Proc. 3210: Mobile Robots XII*, Pittsburgh, 107–118

Kasiński, A. and P. Skrzypczyński, (2002). Communication Mechanism in a Distributed System of Mobile Robots, In: *Distributed Autonomous Robotic Systems 5*, (H. Asama *et al.*, eds.), Springer-Verlag, Tokyo, 51–60.

Nett, E. and S. Schemmer, (2003). Reliable Real-Time Communication in Cooperative Mobile Applications, *IEEE Trans. on Computers*, 52(2), 166–180.

Smith, R. G. (1980). The Contract Net Protocol: High-Level Communication and Control in a Distributed Problem Solver, *IEEE Trans. on Computers*, 29(12), 1104–1113.

Copyright © IFAC Telematics Applications in Automation and Robotics, Espoo, Finland, 2004

ELSEVIER
IFAC
PUBLICATIONS
www.elsevier.com/locate/ifac

MOBILE ROBOTS AND AIRSHIP IN A MULTI-ROBOT TEAM

Jörg Kuhle
Hubert Roth
Niramon Ruangpayoongsak

University of Siegen, Institute of Automatic Control Engineering,
Hölderlinstr. 3, 57068 Siegen, Germany
Phone: 0049-271-740-4439
joerg.kuhle@uni-siegen.de
hubert.roth@uni-siegen.de
niramon.r@uni-siegen.de

Abstract: At the University of Siegen a heterogeneous multi-robot team is created to perform search and rescue or exploration missions. The robot team consists of ground based robots equipped with inertial sensors and 3D vision. The team is assisted by a robot airship flying over the operation field and sharing information with the ground based robots. All robots use a uniform user interface for tele control and additionally supply force feedback manual operator control for human interaction.
Copyright © 2004 IFAC

Keywords: Telecontrol, Mobile Robots, Remote Control, Aerospace

1. INTRODUCTION

This paper describes the actual development of a heterogeneous multi-robot team at the University of Siegen. The robot team consists of ground based robots in a car style and a non-rigid airship, also known as blimp, which explores the environment from air.

Multi-robot teams are of high interest for cooperation and spread capabilities close future application may be search and rescue missions, mine detection and clearing or exploration missions. All these different tasks imply certain distributed autonomy of the robots to survive in the field. To ensure the given target is detected and the mission coordinated, all robots must communicate via radio link. Last but not least all these types of missions require a human control interface, to reply the mission state or even provide assistance by humans. The operator at the mission control center may set search areas, mark possible targets or directly take over control of a robot remotely. At the University of Siegen the operator gets video streams of a web cam, mounted on a robot in the field. To manually control a robot the operator uses a force feedback joystick. By this way direct settings may be given to the robot, additionally the robots gives a kind of "feeling" of the environment back to the operator, like feedback of bumpy grounds and obstacles at ground based robots or wind at the airship. The operator may give autonomy back to the robots at any time.

2. THE GROUND BASED ROBOTS

A series of micro-robots MERLIN has been designed and implemented for a broad spectrum of indoor and outdoor tasks on basis of standardized functional modules like sensors, actuators, communication by radio link. The sensors used for navigation are

- a gyroscope determining the orientation by integrating the angular velocity of the car

- encoders in two wheels estimating the distance and wheel speeds

- ultrasonic sensors for obstacle detection

- bumpers for crash protection

- a 3 axis compass

- for outdoor applications a GPS-sensor working in differential mode

Fig. 1: The MERLIN ground based robot

As shown in Fig. 1, MERLIN is controlled by 80C167 CR 16 bit-processor. The microprocessor is employed for interfacing sensor data acquisition via CAN-bus, sensor data pre-processing, calculation of the control algorithms, and telecommunication with a remote control and monitoring station. The microcontroller, electronic circuits, motors, and sensors are supplied by a 7.2 V and 12 V NiMH batteries. Two motors on MERLIN, steering and driving dc motors, control the direction and the speed of the car.

2.1 MERLIN remote control

MERLIN can be accessed via Internet. A remotely designed path can be downloaded to the robot and executed. The java client and server based on TCP/IP communication are initiated according to the network structure similar to that of the airship COBRA shown in Fig. 6. The two radio transceivers from Radiomatrix are used for wireless data communication between a PC and robot. One is mounted on the robot to the data port of the microcontroller. The other one is connected to a serial port of a server PC.

Two modes of TCP/IP connection are implemented, a local host and a specified IP address host. In the local host mode, the server and client are run on the same PC. In the second mode, two PCs are connected via LAN or Internet. The IP address of the server PC or host must be known and reachable. The client is run on a different PC as a server. In this case, the delay time is bigger than local host mode and must be further investigated. For both of TCP/IP connections, the control and sensor data are transmitted after the server has accepted client. The following sections show the control interfaces via radio transmission.

2.2 Path control

One of the most popular topics in automation is path control, inwhich the path is specified and given to the robot and the robot moves along that path without human interference. The robot has to recognize the path and control its body to move along the specified path and reach the destination. The robot sensors send the signals to the microcontroller and the robot makes a decision and assigns commands to move into the correct orientation. Not only that it is able to stop at the destination but it also moves along the arc of a curve or along a straight line.

The path control in MERLIN robot starts from receiving the drawn path by the user via GUI. The MERLIN server sends that information to the robot and the car performs the path control. The path control panel is shown in Fig. 2. An example of the lines or arcs is also shown. The sensor data, shown on the right column, are from ultrasonic, gyroscope, and 3DM compass sensors.

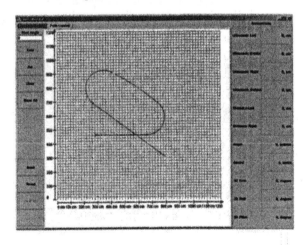

Fig. 2: The path control panel

To increase the data transmission rate, one possiblity is to change from radio transceiver to wireless LAN. PC104 with 933 MHz PentiumIII is selected according to its faster computing performance and abilities to communicate to other devices such as web camera. In our research, the combination of C167 and PC104 is implemented. The serial port transfers the data between them and the Linux operating system is choosed for our application because it is stable and reliable.

2.3 Obstacle avoidance

An intelligent robot must have an ability to protect itself from collision into any objects, which may destroy it in the worst condition. The types of objects depend on the environment. For indoor environment, the obstacles are for example walls, doors, tables and chairs. For outdoor environment, the stones or the tree are the obstacles. The intelligent robot should be able to avoid collision in all cases. Thus, MERLIN detects the object in the front. When the object obstacles its movement, the robot turns into other direction, in which it can further move without collision. Since artificial intelligent technique has robustness and adaptation for problem solving, a fuzzy logic controller has been implemented (Roth, H. et al., September 2003).

In the moment the development focuses on the improvement of environment detection. The use of a normal camera system would constrain the system power down, the image processing needs massive computational power. Thus the environment detection for obstacle avoidance and recognition has to be done by active sensors that provide ready to use data. The technology that provides these capabilities is the Photonic Mixer Device (PMD) technology (Roth, H. et al., April 2003). This device is a real 3D camera.

As shown in Fig. 1, the PMD camera is mounted on the top of the MERLIN robot. The camera measures the distance in the front by using infared sensors constructed on its body. An image of a chair obtained from 16x16 pixels camera is shown in Fig. 3. These data are integrated with the ultrasonic data to improve the performance of fuzzy logic controller for obstacle avoidance.

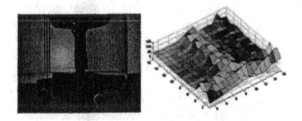

Fig. 3: An image of a chair taken by a digital camera (left) and taken by PMD camera (right)

2.4 Joystick control and haptic interface

In some applications, for example in an unknown environment, an area observation or map creating, where we have no information about the characteristic of surface, it is more powerful to control the robot partly by human for better decision in some situations, where the robot is in danger. Joystick is one of the choices for human interface control. It is a well known plug and play device,

available in the market. Furthermore, the movement direction of Joystick agrees with the movement of the robot. The Microsoft Sidewinder Joystick is selected for our application.

In Joystick control mode, the robot listens to the human commands. The decision to move its body is now up to the joystick movements, forward-backward and left-right. This is sometimes dangerous for the robot when the user cannot recognize obstacles but he still keeps moving the joystick into the front. The robot may crash into the obstacle and destroy itself. For intelligent robot, when it sees obstacles, it should automatically obey the command from the joystick and try to avoid collision. When it is safe from collision, it should continue performing the Joystick command. The joystick control panel is shown in Fig. 4. The angle of rotation in roll, pitch, and yaw directions are also shown in the car model. From these values, the top and side views of robot from far distance are also shown on top left and right of the panel. These car images represent how the car is turning or tilting on the ground surface.

Two modes of operation are implemented, joystick alone control and joystick with obstacle avoidance control. In the first mode, MERLIN always performs the command from the joystick even if it founds the obstacle.

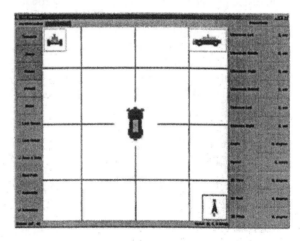

Fig. 4: The joystick control panel

In the second mode, MERLIN listens to the joystick commands until it founds some obstacles. Then it obeys the joystick command and uses fuzzy logic controller to avoid collission. The fuzzy logic controller is stoped when there are no more obstacles and the car continues performing the joystick commands.

Sensing or force feedback joystick is another interesting topic for mobile robot. What will happen when a user can feel what a robot feels? The haptic interface is included to realize the feeling of the robot when it is moving. Like an usual car, when it moves fast, it feels some air resistance, which resist it body to move forward and it must supply more force to the wheels, i.e. the higher velocity, the more air resistance force. Joystick represents that force by creating force feedback on its body. The user can feel some force on his hand when he tries to move the joystick forward. That means when he wants to move into the front faster, he has to put more force on the joystick, too.

3. THE AIRSHIP

At the University of Siegen a model airship is constructed to operate as flying platform. Flying robots give a third dimension to normal ground based robots (Newman, 2002). The idea of creating such a robot is to enhance the capabilities of a heterogeneous multi-robot team. The airship robot was created to observe the environment and detect possible targets, which may be handed over to the mobile ground based robots. Thus the airship is called COBRA for "Cooperative Observation Robot Airship". The airship has a size of around 3m length and 80 cm diameter and is used indoor. For outdoor use a larger model will be created later, and control mechanisms of real airships might be applicable (Hagenlocher 1999, Roth and Adermann 1994/1995).

3.1 Position control

On of the first steps when creating a mobile robot is to implement position control and its sensor equipment. Certainly measuring the position in the air is quite complicated and inaccurate compared to measuring on ground. So COBRA got a three-dimensional gyroscope and low-gravity accelerometers to detect any movement. The sensors are mounted on a special circuit called BeeCon Bluewand at the middle of the hull and send the data via Bluetooth connection to mission control. At this robot control server the data is integrated to gather position, speed and angle of the airship. This information is send back to COBRA via radio link and considered into local control algorithms on the microcontroller. Although these sensors give good results, the measurement tends to drift after a while, so the robot must get some more, reliably data. The airship is build up with low-cost model plane motors and has a size of 3m length, so the airship is used indoor at present, therefore no GPS receiver for absolute position measurement may be used. But an indoor absolute position system called V-Scope is used in tests at present. Any triangulation or time-of-flight system might enhance the capabilities of the sensors. The V-Scope uses ultrasonic to detect small markers at the airship and gives all information to mission control. The airship is equipped with small markers, which can be detected by the V-Scope system, mounted on the ground. It uses time-of-flight of the sonar signals to detect the distance and infrared detectable colours on the markers to differ them. The only negative aspect is the maximum range of 5m.

But it can be used as basis to correct the absolute position from time to time. In fact unreliable data may be the only information available in the field.

Fig. 5: The airship COBRA

COBRA should also be equipped with an ultrasonic or infrared sensor for relative height measurement for landing, but it is not implemented, yet. It gets an absolute height from a pressure sensor mounted on the ship. This information gives the robot its capability to control standstill for now. The resolution of the pressure sensor supports height differences of approx. 50cm. When in use, the airship is flying for some minutes, so no weather changes must be considered for the height control. But for later use a similar circuit is used as reverence at mission control. The airship is shown in Fig. 5 flying at the university site.

3.2 Operator control

To achieve the goal of a rescue or exploration mission, the airship can also be controlled by a human operator. The control is done by a force-feedback joystick, the same as for the MERLIN robot. The manual control overrides the autonomous control of the airship. In this mode COBRA will stabilize the position after release of the joystick control. As there is no obstacle avoidance or autonomous guiding system for the airship at present, the operator gets full control of the airship while taking over. The idea for this mode is to manoeuvre the ship quickly into its starting or target position. The sidewinder joystick gives many ways to implement 3 axes control needed for the airship. Even if the control of a 3D flight might be tricky using a joystick, all robots should be able to use the same software and interfaces for human interaction. This allows an easy integration of new robots and ensures the usability for the operator.

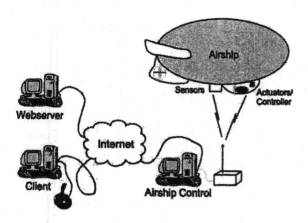

Fig. 6: COBRA network configuration

The network configuration is shown in Fig. 6. The operator will work at the client site and launch the application of the web-server. The airship control server will send data from camera and sensors to the client. The communication to the airship is done by serial radio link, while the inertial sensor system sends its data by Bluetooth connection to the server.

4. ROBOT TEAMWORK

The actual simulated scenario is a search or exploration mission for the multi-robot team. The COBRA airship is controlled using a force feedback joystick by the operator. While moving, the airship logs its position and gives feedback to the mission control. As the operator spots a possible target, the control will be given back to the autonomous COBRA control. The airship is now holding position over the target itself and asks the mission control for sending out exploration ground based robots over network. At this moment the MERLIN robots swarm out, using the given position of the airship and their own autonomous driving. Obstacles are detected and possible paths driven automatically by the robots. When reaching the target, the robots give feedback to mission control, i.e. task completed. In a real mission the robots should examine the target. For the actual simulation step every target counts as positive detection when found. The exploration of ground based robots with the airship controlling the target position is shown in Fig. 7.

The MERLIN ground based robots use 3D-Vision and ultrasonic sensors for obstacle avoidance. The path is generated automatically by setting target position. The movement of the robots can be supervised by a Java application and video stream. All robots give back speed, direction, position and obstacle information. The airship is moving around the given target position. After reaching the target the robots send a task complete command to mission control and may be used for another task.

Fig. 7: Cooperative exploration mission

For an easy use of all robots, a uniform user interface is developed for joystick control and position feedback. The different robot types can be set up in the user interface, so the joystick and simulation parameters are adapted to the actual robot, i.e. the joystick may get a third axis for airship control or the interface provides different information. By this way even new robots can easily be added to the heterogeneous robot team. The communication over serial radio link uses a uniform protocol which is used by all the robots. Robots can be identified by a significant name and number. When adding a robot to the team, the identification will pop up in the user interface.

Even if the robot interface is uniform for the client, the hardware configuration of all robots is different. The MERLIN is equipped with large computational power, including a PC104 computer, while the airship just uses a C167 microcontroller, because it can only carry lightweight controllers, sensors and power supply. Thus the computational power is given by the COBRA airship server. This server is located in reach of the radio link of the airship. Thus the server also gives the possibility to the user to get airship data and enable more complex control. The server may use MATLAB/Simulink with real-time hardware to compare the real plant with a model of the airship while running. The MERLIN server on the other hand just has to get mission definitions and to generate single tasks, e.g. define a new target. The robot will perform more autonomously. To combine all the different systems, the robot servers have to provide uniform interfaces and operate the robot as "black box" for clients. The communication protocol to the robot servers is based on mission definitions, e.g. clear the room or search the target. The servers divide the missions into single tasks or steps and control the robots itself. The client just gets mission states and can interact with the servers while operating. The configuration is shown in Fig. 8.

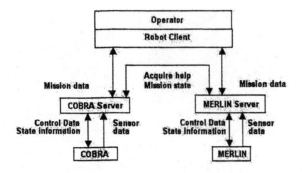

Fig. 8: Robot communication diagram

5. CONCLUSION

The multi-robot team can survive in simulated mission, i.e. a laboratory surrounding with clearly defined targets. The airship enhances the search capabilities of the robots. 3D vision is a good way for obstacle avoidance. In the future the airship should be able to map the environment and even use this map for navigation. This high computation power can be provided by the mission control server.

Haptic control gives a kind of "being online" feeling to the operator. But this feedback should contain obstacle, bumpy floor, etc. information to give a real feeling of the environment. But actually the use of joysticks and manual control can often improve the success of a robot team. It combines the spread capabilities of many robots with a human brain and experience.

Telematics gives small robots with limitations in payload and volume the possibility to shift computational power and control to an external host. It also provides users with more detailed information about the system. In the robot team presented in this paper the user may take over complete control or give autonomy to the robots and just define targets.

6. REFERENCES

Hagenlocher, K.G. (1999). A Zeppelin for the 21st Century. *Scientific American*, (http://www.sciam.com/1999/1199issue/1199hagenlocher.html).

Newman, D. (2002). *Interactive Aerospace Engineering and Design*. Boston: McGraw Hill.

Roth, H.; Adermann, H.-J. (1994). *Flight Control Concept of the Zeppelin NT Airship*, Proc. "ICSE '94", 10th International Conference on Systems Engineering, Coventry, UK, September 1994

Roth, H.; Adermann, H.-J. (1995). *Airship Control using Adaptive Fuzzy Logic Strategies*, Proceedings of the 6th International Fuzzy Systems Association World Congress, Sao Paulo, Brasilia, p. 683-686, 1995

Roth, H.; Schwarte, R.; Ruangpayoongsak, N.; Kuhle, J.; Albrecht, M.; Grothof, M.; Heß, H. (2003). *3D Vision Based on PMD-Technology for Mobile Robots*, Aerosense - Technologies and Systems for Defense & Security 2003, SPIE Conference, Orlando, Florida, April 2003

Roth, H. et al, (2003). 3D Vision Based on PMD-Technology and Fuzzy logic control for Mobile Robots. *Second International Conference on Soft Computing, Computing with Words and Perceptions in System Analysis, Decision and Contro*, Antalya, Turkey, September 2003

Copyright © IFAC Telematics Applications in Automation and Robotics, Espoo, Finland, 2004

MULTIROBOT SYSTEM ARCHITECTURE & PERFOMANCE SSUES FOR THE UJI ROBOTICS TELELAB

R. Marín P. J. Sanz P. Nebot R. Esteller R. Wirz

Universitat Jaume I
Av. Sos Baynat, s/n. E-12006 Castelló - SPAIN
{rmarin, sanzp, al065289, esteller, al003904}@uji.es

Abstract – In order to design a Robotics Tele-Laboratory that permits simultaneous remote control and programming of several robots, we must take into account multiple aspects. One of the most important refers to the System Architecture and Communication Technology that permits such a level of multirobot interaction. In this paper we present the system architecture that has been used in our laboratory to let users program remotely two educational and two industrial robots through the same web-based system. Experimental results show different alternatives to organize the architecture and focus on the system performance aspects. *Copyright © 2004 IFAC*

Keywords: Telemanipulation, TCP/IP distributed systems, Robotics TeleLabs, Remote Programming, and Education & Training.

1. INTRODUCTION

In October 2000 the UJI Online robot was connected to the web for research purposes. It consisted of an educational robot (Mentor), with three cameras that enabled a user to remotely control pick and place operations of objects located on a board. Experiments about object recognition, virtual reality, augmented reality, speech recognition, and telemanipulation were accomplished in order to enhance the way people interacted with the system.

Since then, we realized the experience would be very interesting in the education and training domain. The first experience in education and training with students was performed in November 2001, were they programmed several pick and place operations on the virtual environment. This was very interesting for them because they obtained a more practical view of the Robotics subject, which until that moment was practiced in a more theoretical manner. After that, they did some manipulations using other online robots (e.g. Telegarden, Australia's Telerobot, etc.), letting them to compare different ways to design user interfaces for telerobotic systems. In November 2002 the second experience in education and training came up. Again, we provided the possibility to the students to program more sophisticated pick and place operations, using this time both, the off-line and the on-line robot.

After that, we considered necessary to enhance the system in order to not only let students control the robot from a user interface, but also programming it by using any standard programming language. In April 2003 the UJI TeleLab project came up, which supposed the design of a Java library called "Experiments" that let students to program their own control algorithms from any computer connected to the internet and execute

them over the real robot. Some pilot experiments were performed with researchers and students since then. The interest for the design of Internet-based Tele-Laboratories is increasing enormously, and this technique is still very new. A very good example of already existing experiments in this area can be found in (McKee,2002). At the same time, we were very concerned about extending the project to letting users to program not only one educational robot, but also several manipulators. In fact, at the moment of writing, the UJI TeleLab lets scientists and students program not only one educational robot from home, but also design algorithms for two industrial and two educational manipulators.

2. EXPERIMENTAL SETUP

As appreciated in Figure 1, we have 4 robotic systems:

(1) Two educational Mentor robots (Figure 1 first row) where seven cameras are presented: two taking images from the top of each scene, one pantilt camera from the side, two cameras situated in two grips and two more cameras from the front. The top cameras are calibrated, and used as input to the automatic object recognition module and 3D-model construction. The other five cameras (two cameras each Mentor plus a one pantilt camera to both) give different points of view to the user when a teleoperation mode is necessary in order to accomplish a difficult task (e.g. manipulating overlapped objects).

(2) One industrial Adept One robot (see Figure 1 second row) operating on top of a conveyor belt. A static camera on top of the scenario provides a calibrated view of the objects.

(3) One industrial and redundant PA10 manipulator (see Figure 1 third row) placed on top of a Nomadic mobile robot.

(4) One industrial Motoman SV3X is as well integrated into the Telerobotic Framework.

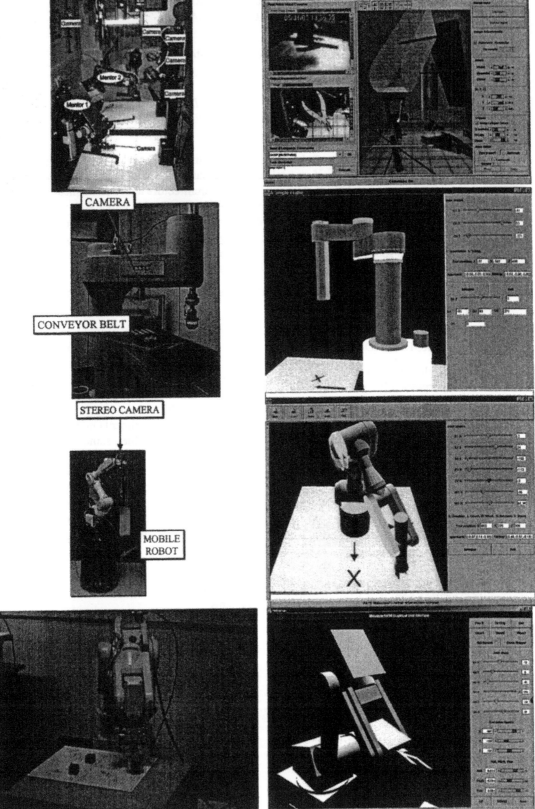

Figure 1. Experimental setup: Multirobot configuration

Once the user gets an online connection to a robot, the manipulator goes to the initial position, so the whole scene information is accessible from the top camera (calibrated position). At that moment, the computer vision module calculates a set of mathematical features that identify every object on the board (Marin, 2002).

Afterwards, the contour information is used in order to determine every stable grasp associated with each object (Sanz,1998), which is necessary for vision-based autonomous manipulation. This process is a must whether high level task specification from a user is required, and also in order to diminish network latency.

Meanwhile, we shall use the output from these visual algorithms in order to construct a complete 3D model of the robot scenario (i.e. the objects and the robot arm). Thus, users will be able to interact with the model in a predictive manner, and then, once the operator confirms the task, the real robot will execute the action (Marin, 2002).

3. MULTIROBOT HW ARCHITECTURE

In previous projects, we have experimented different ways of accessing a single robot by several remote users (Marin, 2002). A possibility would be letting just one operator to have control on the real robot, while the others are programming an off-line virtual environment. As an alternative to this problem (i.e. having a unique robot and several users), a possible solution would be allowing access to more than one robot at the same time (see Figure 2).

The user selects in the user interface the robot he/she wants to interact with. The result is having several operators that are using a maximum of four robots at the same time, while the others are experimenting with simulated off-line virtual environments. In any case, while somebody is performing a manipulation in any of the situations abovementioned, he/she could ask other people connected into the system to help (via chat) in a particularly difficult task.

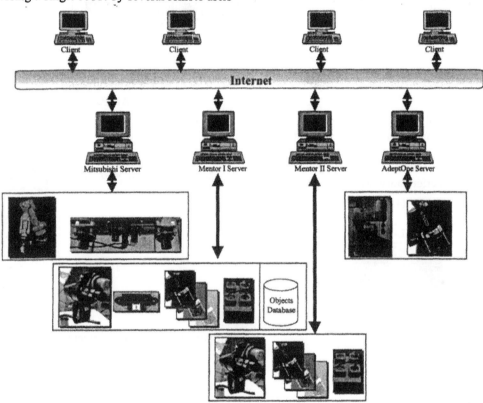

Figure 2. Multirobot HW architecture

4. MULTIROBOT SW ARCHITECTURE

As we can see in Figure 3, the Tele-Lab accepts experiments (i.e. algorithms) as inputs using any programming language capable of managing TCP/IP sockets. We already provide a Java library for using the robot in a simply manner.

The outputs of the experiments are returned to the user by means of the Tele-Laboratory "Client Side" user interface, which permits the operator to see the results of their programmed actions directly on a multimedia Java3D user interface. The client side applet launches the user interface that allows the user to interact with the remote robots. The connection of this applet with the server side (Servers Manager module) is performed through a unique connection via RMI (Java Remote Method Invocation API). It means every connection for every one of the robots is passing through this server. Of course, by having many robots connected to the system the Servers Manager would be a bottleneck. However, this configuration is very important for the synchronization of the robots' operations as well as for the specification of reliable multirobot tasks. The Servers Manager connects to the different robot servers by using the CORBA standard. It facilitates enormously the configuration and installation of the different server sides (robots' servers).

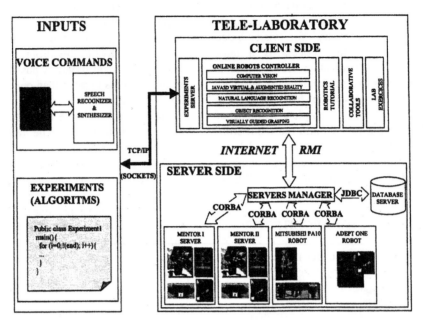

Figure 3. Software architecture for the Internet-based Tele-Lab

5. RESULTS

In this section different configurations of the architecture are presented. For every configuration we have performed 400 remote movements on the robot. Some experiments have been done by using TCP, UDP and RMI protocols. Performance details are given.

5.1. UNIQUE CLIENT/SERVER IP (UCS_IP)

Figure 4. Unique Client/Server IP (UCS_IP) Software Architecture

By looking at Figure 6 we can appreciate that the latency **on campus** for both, TCP and UDP protocols is convenient enough to perform teleoperation experiments. Although there are some robot movements that have invested almost 14 msecs on communicating with the server side, the average is less than one msecs. Moreover, by looking at Figure 7 can be seen that by teleoperating the robot from home (same city) using the TCP/IP approach is rapid enough for our purposes (less than 90 msecs of average).

5.2. TELELAB EXPERIMENT RMI (TE_RMI)

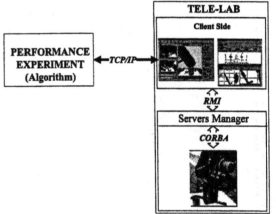

Figure 5. TeleLab Experiment RMI (TE_RMI) Software Architecture

The second configuration consists of using the TeleLab feature in order for the students and scientists to program their own algorithms remotely. This means the experiment will connect first via TCP/IP on the local machine with the TeleLab user interface (Client Side), and then, once the operation is confirmed, the action is executed on the real robot via a RMI connection with the Servers Manager. Again, the Servers Manager connects via CORBA with the remote robot.

It must be taken into account that once an operation is sent to the Client Side, this operation is executed virtually over the predictive interface, and then, once confirmed the action by sending the command "confirm", this operation goes through the RMI and CORBA interfaces to move the real robot. It means this "predictive" configuration is very useful for students that are working off-line or just want to make sure on the virtual environment about the next robot movement.

And of course, it supposes the time latency is increased a lot. In fact, the average for the experiments executed on campus is 2227 msecs, and for those working from home is 2439 msecs.

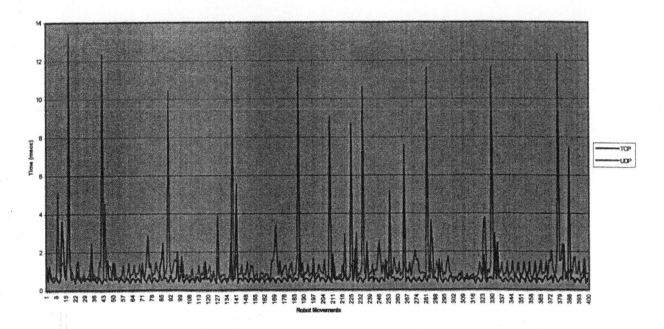

Figure 6. Graphical representation of the latency for the configuration Unique Client/Server IP (UCS_IP) **on Campus,** using both, the TCP and the UDP protocols

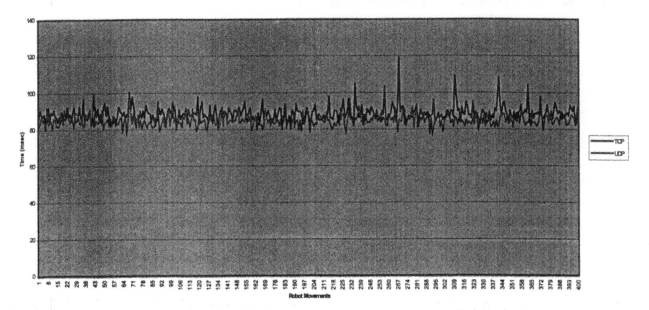

Figure 7. Graphical representation of the latency for the configuration Unique Client/Server IP (UCS_IP) **in the same City,** using both, the TCP and the UDP protocols

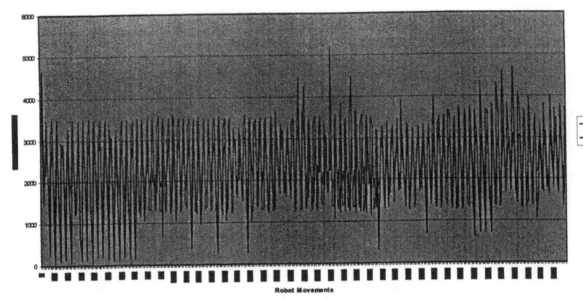

Figure 8. Graphical representation of the latency for the configuration TeleLab Experiment RMI (TL_RMI), using both on Campus and in the Same City configurations

6. CONCLUSIONS

The present paper has presented a pilot experiment with several telerobotic configurations that enable students and scientist take control of a robot remotely. The first configuration (UCS_IP) is the fastest one, and is the most convenient for traditional telerobotic applications. In fact the latencies shown in Table I justify this.

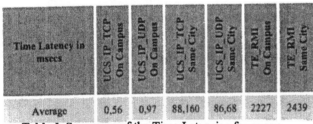

Time Latency in msecs	UCS_IP_TCP On Campus	UCS_IP_UDP On Campus	UCS_IP_TCP Same City	UCS_IP_UDP Same City	TE_RMI On Campus	TE_RMI Same City
Average	0,56	0,97	88,160	86,68	2227	2439

Table I. Summary of the Time Latencies for every configuration experimented

On the other hand, we have presented the Tele-Lab Multirobot Architecture (RE_RMI) that permits any student or scientist to not only control the robot from a user interface, but also allows him to control the manipulator from a Java program (remote programming). This second configuration is much more expensive because it requires a predictive display feature (requires confirmation of the command) and also is uses not only a TCP/IP protocol but three (TCP; RMI and CORBA). As a pilot experiment we consider the results obtained with the TeleLab configuration are good enough for several kinds of applications (e.g. Pick & Place experiments). By the other hand, performance should be improved a lot if more sophisticated experiments are going to be implemented in the future (i.e. Remote Visual Servoing).

ACKNOWLEDGMENTS

This is to acknowledge sources of financial support for this research, that was provided in part by the Spanish Ministry of Science and Technology (CICYT) under project CICYT-DPI2001-3801, by the "Conselleria de Cultura i Educació" (Generalitat Valenciana, Spain) under project GV01-244, and by the Fundació Bancaixa under projects P1-1A2003-10 and 031290.01/1

REFERENCES

Marín R., P.J. Sanz., J.S. Sanchez (2002) A Very High Level Interface to Teleoperate a Robot via Web including Augmented Reality. In Proc. of the IEEE Int. Conf. on Robotics and Automation (ICRA). Washington.

Marín R., J.S. Sanchez, P.J. Sanz. (2002) "Object Recognition and Incremental Learning Algorithms for a Web-based Telerobotic System". In Proc. IEEE Intl. Conf. on Robotics and Automation (ICRA). Washington.

McKee G. T. (2002) The Development of Internet-Based Laboratory Environments For Teaching Robotics and Aritificial Inteligence. In Proc. of the IEEE Int. Conf. on Robotics and Automation (ICRA). Washington.

Sanz P.J., A. P. del Pobil, J. M. Iñesta, G. Recatalá. (1998) "Vision-Guided Grasping of Unknown Objects for Service Robots". In Proc. IEEE Intl. Conf. on Robotics and Automation (ICRA), pp. 3018-3025, Leuven, Belgium.

Copyright © IFAC Telematics Applications in Automation and Robotics, Espoo, Finland, 2004

ELSEVIER
IFAC
PUBLICATIONS
www.elsevier.com/locate/ifac

ROBOT/HUMAN INTERFACES FOR RESCUE TEAMS

Frauke Driewer, Herbert Baier, Klaus Schilling

Julius-Maximilians-University Würzburg, Informatics VII: Robotics and Telematics
Am Hubland, 97074 Würzburg, Germany
{driewer, baier, schi}@informatik.uni-wuerzburg.de

Abstract: Mobile robots can be of significant support for human teams in search and rescue situations. A user requirement analysis on basis of questionnaires distributed to fire fighters and emergency support people expresses notable interest in sending teleoperated robots into dangerous areas instead of risking human life. This paper summarises the requirements and the consequences for a telematic system with an intuitive user interface, which supports quick reaction capabilities. Design and implementation of an infrastructure addressing data flow in joint teams of humans and robots for data sharing, teleoperations and remote coordination are presented. Furthermore, robot vehicles used in that context and test scenarios are described. *Copyright © 2004 IFAC*

Keywords: teleoperation, telepresence, mobile robots, user interfaces, man-machine interaction, cooperation

1. INTRODUCTION

Mobile robots can assist or even replace humans in dangerous and hazardous situations. Typical application areas emerge in space, under water or in other harsh environments, where access is impossible or very dangerous for humans. The use of mobile robots for search and rescue services during catastrophes and disasters represents such a scenario. Robots with appropriate equipment can take sensor measurements to characterize the environment and localize themselves also in situations adverse for humans, such as at low-visibility conditions. Nevertheless, robots can only complement the humans with their far superior adaptive reaction capabilities. In order to take advantage of these complementary capabilities, telematics techniques can be used to support joint teams of humans and robots by means of data sharing, teleoperation and remote coordination.

Tasks for mobile vehicles are the exploration of unknown regions and the search for victims. During their mission the robot/human rescue team is supported by mission coordinators. They are located at a safe remote site (outside the region of accident) taking advantage of a well functioning infrastructure.

The support provided by the remote coordinator varies from guidance of the team to provision of additional background knowledge about the environment and the situation. The team members are equipped with sensors for localization and environment characterization. On basis of these measurements the coordinators know the position and status of all team members, as well as environmental conditions. Thus, an overview of the situation can be obtained that allows improved planning of tasks. The coordinators guide the human team members through the environment and command the mobile robots. This is supported by algorithms for efficient cooperative coverage in searching the complete area and by a user interface, which provides all the necessary information in an intuitive way, enabling quick reactions in time critical situations.

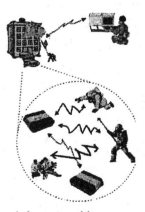

Fig. 1. Human/robot team with remote coordinator.

During discussions with professionals in search and rescue, such as fire fighters, it was obvious that there is a wide interest in the application of robots for emergency cases. Nevertheless, there exist also concerns against the use of new, possibly immature technology. Therefore, a careful investigation of the user's needs and requirements is absolutely essential for the development of rescue systems.

This paper describes the requirements analyses based on questionnaires distributed to emergency support groups as well as on detailed interviews with fire departments and other institutions, dealing with emergencies. According to the user requirements a typical scenario was derived, robots were selected for tests and the basic system design for a human/robot rescue team supported by a remote coordinator is presented in this contribution.

The presented work is part of the PeLoTe project (Building *Pre*sence through *Lo*calization for Hybrid *Te*lematic Teams) funded by the European Community within the "Future and Emerging Technologies" program targeted on "presence" methods and applications.

2. USER REQUIREMENTS

The user requirements are derived from answers to questionnaires sent to fire departments (Germany, Finland and Czech Republic) and governmental disaster relief organizations (Germany). Different types of fire brigades, related to cities, airports and industrial plants, replied to the questionnaire. The proposed cooperating human/robot system seems to be especially interesting for professional fire fighters, e.g. in chemical factories or airports, or for rescue cases in big office buildings, as well as after emergencies with massive destruction, like for earthquakes.

For modern buildings there are usually maps of the building available as well as additional information, e.g. storage areas of dangerous materials or positions of activated fire alarm sensors. Due to the progressing destruction caused by the emergency, this map might be continuously changing. In order to represent all the knowledge about the environment, the building map will be combined with additional information in a layer-based electronic format. This is available for the remote coordinators, as well as for the humans in the emergency area and will be continuously updated. In that way sharing of up-to-date data between team members is possible and allows quick and meaningful warnings, when unexpected situations arise.

The layer-based format (cf. Fig. 2), which allows display of currently required information, received good acceptance, since maps used nowadays contain often too much information at once. Information that was considered as useful, are the ground plan with

stairs doors, windows, rooms, access routes, valves, pipelines, shut-off socks, energy systems, hazardous material and all fire fighting relevant objects, such as possible position of fire source, fire reporting centre, exits and entry ways, extinguish systems and sources of water. During the mission other data will be added dynamically to this map, such as the position of rescue personnel and victims or polluted areas. The requested accuracy for position of robots and humans, including victims, is about one meter and the required update time of information should be in the range between one and ten seconds. For dangerous materials the specification of the region is sufficient. Highly demanded was also the visualization and supervision of the breathing protection system. Most preferable way to display map information to the rescue personnel in place is a head-mounted display, followed by an arm-mounted display.

The robots are considered to be especially useful for exploration of an unknown emergency area and search for victims. Currently, telerobots are preferred instead of fully autonomous robots. However, autonomous features supporting operations like path planning with obstacle avoidance and person following capabilities are desired. For future developments, fire fighters like to have autonomous robots that are able to accomplish tasks, such as searching wide areas for victims and exploration of dangerous areas on their own. Although, they are in some cases cautious about the reliability and adaptability of autonomous vehicles. Nevertheless, the employed telerobots are provided with autonomous low-level reactions and emergency handling situations, satisfying high criteria for robustness. Good mobility is an important feature for the robot, in order to take measurements in areas, which are difficult to access. Data acquisition from these areas is a crucial point. Camera pictures from the scene are required from almost all rescue personnel. While audio communication, which permits the rescue personnel to speak with found persons through the robot, was required from half of the participants. The robot design should exhibit a high robustness to extreme environment conditions, such as little oxygen, explosive atmosphere, high or low temperature regions.

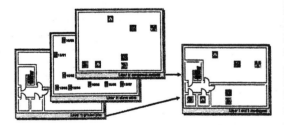

Fig. 2. Example for layer-based rescue map: Building map overlaid with additional information on fire sensors and storage sites for dangerous material.

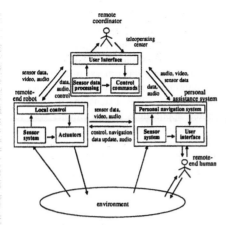

Fig. 3. Schematic of information flow and processing in the human/robot team in place and the remote coordinators.

The techniques developed in the PeLoTe project receive good acceptance from the rescue community. There is great interest in localization for the team members during the mission. The general concept of telematic issues (telerobots and remote coordinator) for rescue scenarios has been confirmed by the users. Robust communication in an emergency case is crucial for the end-user system. In order to meet this criteria, strategies are needed that deal also with interrupted communication, as autonomous features for the robots and backup communication. In order to deploy the layer-based map, an informative, but intuitive user interface is necessary. This also includes methods to display the camera pictures and properties of the team members, as well as to broadcast information, message and warnings. Efficiency in personnel placement and task sharing was demanded, which will be provided by cooperation and coordination strategies.

3. SYSTEM DESIGN

This section addresses a generic cooperative system composed of humans, robots and remote coordinators for handling a given rescue task. In the special case of a search and rescue scenario, the environment is quite hostile. The information sharing between the team members needs to be very quick and efficient, since actions are time-critical and the humans in place (rescue personnel and victims) are in danger. A well-arranged user interface should provide a complete overview of the situation. The visualized information has to be presented in an efficient and easily interpretable way, such that the coordinator can react quickly and in a right manner to the emergency situation. The humans in place need a simplified user interface and a personal navigation system (Saarinen et al., 2004). Furthermore, the developed system should not give additional load to the fire fighters. Thus, tasks regarding the infrastructure of the system, as e.g. establishing communication, are mostly executed automatically by the software.

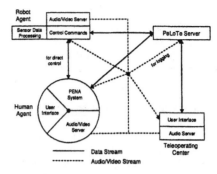

Fig. 4. Client/server system architecture.

Fig. 3 summarizes the flow of information and control between the remote coordinator and the team members in place. Video cameras and audio transmission devices provide a realistic impression about the situation in the emergency site without being there. The human team members and the remote coordinator are using audio transmission for their standard conversations, e.g. for the exchange of observations. The robots carry also audio transmission devices in order to work as communication platform to the found persons. Other kinds of information are sensor and navigation data, as well as control commands. In order to achieve this system design two software architectures have been considered: client/server and event based architectures.

Fig. 4 displays the principle of the client/server architecture, a centralized architecture. All data communication passes through the server, which also stores these data for documentation uses. The server only responds to requests from the clients and it will never send an update to a client without a previous request. Thus, the clients need to ask for updates. In such architecture, the server does not take care about the communication. If it breaks, the concerned client has to reestablish it. When a client need some periodically update, it should pool the server. The clients can be divided into following categories: user interfaces for the remote coordinators (teleoperating centre) and agents for the humans and robots in the team (environment). Exceptionally from the centralized architecture audio and video can be requested directly by the clients. These data do not necessarily pass through the server. Furthermore, a human member in place can control a robot directly.

The human team members have their personal navigation system and a user interface similar to the one in the teleoperating center. The software of the robot team members includes sensor data processing and navigation control. Each client should register itself in the PeLoTe server before it can participate in the system. Hence, the server is able to trace and to spread information from clients.

An alternative to the client/server architecture offers the event based inter-process communication, which is decentralized. The event traffic is managed by a postmaster background process (*pmd*). It provides event handling, forwarding, and message passing between processes. When *pmd* receives an event, it forwards it to all the processes that are registered to receive this event. The global structure is shown in Fig. 5.

Postslave processes (*psd*) can be used to extend the system. Each post process (*pmd* and *psd*s) can monitor a subset of the remaining processes, like human and robot agents. The post processes are organized hierarchically, where the *pmd* is the master. There are two types of event flow between the post processes, hierarchical and cooperative forwarding (cf. Fig. 6). Hierarchical forwarding describes the direct communication between *pmd* and *psd*. The cooperative forwarding uses another *psd* as bridge and can be used in the case of communication deficit. This feature provides not only extension, but also robustness against communication failures and allows local communication between processes monitored by a post process.

Both architectures show strengths and weaknesses for the search and rescue application. In the event based architecture the post process handles the communication and therefore also its troubles, e.g. reestablishing of connections. This is not the case in the client/server architecture, since the server only responses to requests of the clients, which are themselves responsible for their connection. On the other hand, the continuous requests from the client side lead to unnecessary network traffic, if no new information is available.

The complete configuration of the system is automatically retained in the server. The event based architecture requires the integration of a special process that carries out this task (*confmang*). All events need to pass this process to keep track of the actual configuration, which means increase of network traffic. The extension to wide-spread distributed applications is in a simpler way realized by the event based architecture. However, its general implementation is very complex, since all event flows must be exactly defined. An event can generate subsequent chains of events.

For the PeLoTe project the client/server architecture was chosen. A server was implemented and the integration of different modules is going on. The server is the main component for data sharing in the PeLoTe system and supports multitasking and multi-user capabilities as well as portability (platform independent). The server has a modular structure, with a hierarchical communication between the modules, which are organized in levels (cf. Fig. 7). The Kernel is the only module of the first level. Each of the modules, except the Kernel, communicates with exactly one other module from a level one step below. Communication between modules of the same level is not allowed. A module can communicate with any higher level modules. Extension of the server is straightforward with this modular hierarchical concept: If a new module should be included, only an interface needs to be provided by a lower level module.

The core of the server is the Kernel module, which is the central point for data sharing and control. It is responsible for the configuration management (manages the current status and configuration of the team members and environment), persistence (creates and maintains log files and save the configuration of the system) as well as authentication and authorization of clients. The application programming interface (API) module allows third party software to communicate locally with the server. An extension of the API is the RMI (Remote Method Invocation) module based on Java RMI, which enables remote communication. Thus, a client can call a remote object in the server in an easy and standard way.

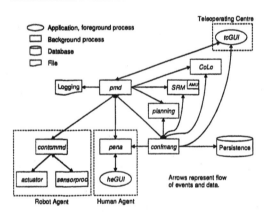

Fig. 5. Event based architecture.

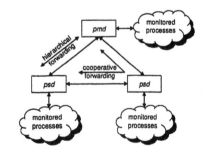

Fig. 6. Extension of event based architecture.

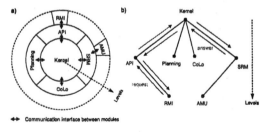

Fig. 7. a) Server modular architecture. b) Example: RMI modules request information from SRM module. The communication goes through API and Kernel modules.

So far, the following modules are included:
- Planning: Responsible for the path planning of the multiple entities.
- CoLo: Performs the cooperative localization tasks.
- SRM: Deals with the tasks concerning the standard rescue map. The AMU module is an extension of SRM for the automated map update.

In order to increase the usability of the PeLoTe system a graphical user interface (GUI) is designed. The GUI for the remote coordinator and for the human team members includes a two-dimensional map as well as the team member properties and data (sensor data, video). The remote coordinator can observe the environment and the team, update the map, command the robots and guide the humans from a remote place outside the emergency area. The GUI and especially the two dimensional map provides also a communication platform. Together with the audio communication and the video images, it generates a telepresence scenario increasing the impression of the concerned persons to be in the same place.

From the standards of user interface design (Wessel, 2002) and the user requirement analyses, guidelines for the man-machine interfaces have been compiled:

- Feedback to user: The user is informed about the status of the system and the team members, e.g. the user receives alarms related to communication interrupts. Delay times are kept as small as possible. The system has multitasking abilities.
- User control: Background tasks are automated and invisible for the end-user, e.g. regular update of the position of the team members. However, each user has full control over his/her own system.

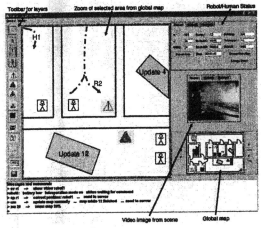

Fig. 8. User interface design: The zoom window shows positions of human 1, robot 2 and found victims, as well as dangerous areas and map updates.

- User's language: The GUI uses the terminology used by firefighters.
- User's logic: Testing and evaluation by qualified test users will be performed to provide feedback to the programmers.
- User group: For designing the GUI, a user group is considered that is familiar with technical equipment, but most probably not an everyday user of computers. Before using the equipment in emergencies significant training practice has to be passed, so novice users are not considered.
- Standardization: Common code of practice, e.g. color or shapes is used. Icons and pictograms that are standard in the fire fighting community are utilized.

4. TEST SCENARIOS

Reflecting the user requirements, test scenarios have been developed, which allows the evaluation of the design system. The experiments will address an indoor environment, which represents a typical large office building. The characteristics of these test fields are: many rooms, irregular shape and dead ends. Parameters as visibility, number of victims or configuration of the environment (e.g. source of fire, dangerous areas) will be changed. Moreover, different stages of difficulty can be arranged (Murphy et al., 2000), e.g. the victims are easy to find or can be hidden behind objects. The tests will be performed by traditional human-only and robot/human teams. From the direct comparison, the benefits of the use of robots, as well as drawbacks and points for further improvements will be identified. Assessment factors are e.g. the time needed to complete the task by different teams and the safety for the rescue team. A more detailed evaluation will include the test observation by emergency professionals, documented in interviews directly after the test.

5. RESCUE ROBOTS

As test vehicle the MERLIN (Mobile Experiment Robot for Locomotion and Intelligent Navigation) robot (Schilling and Meng, 2002) is used for the search and rescue scenario. The length is about 40 cm. It is equipped with a C167 microcontroller and radio link (Radio Packet Controller). The motors and the board are supplied by battery. The rover is equipped with ultrasonic and odometry sensors and performs autonomous low-level actions, such as collision avoidance.

In order to provide further processing capacity and handle WLAN access, a PC 104 board is connected to the microcontroller via serial port. Due to its CAN-bus interface the rover provides good flexibility and an excellent test bed for new sensors, control methods and algorithms.

There exist tracked and wheeled versions to offer suitable vehicles for a broad range of application scenarios. The tracked version of MERLIN provides the same features and is equipped with similar sensors, but different locomotion properties. The tracked wheels are more suitable for moving in destructed areas, but lead to increased power consumption and therefore to a reduced operations period.

In order to relieve the operator, further autonomy features are implemented, as execution of movement with obstacle avoidance. This prevents the operator from the task of continuously teleoperating the robot. Another feature of the robot is the autonomous pursuit of a person. This is useful in order to send a small human-robot team as a group to the same target. In this context work related to convoy driving and cooperative navigation has been analysed with the MERLIN robots (Gilioli and Schilling, 2003). This work concerns the development of navigation algorithms for multiple mobile robots on basis of ultrasonic range measurements.

6. CONCLUSIONS

From the evaluation of user requirements by questionnaires, an architecture for a heterogeneous human/robot search and rescue systems was derived, including in particular telepresence methods for remote coordinators. Generic techniques for cooperation of humans and robots in joint groups have been developed and related software systems are implemented. The core of the server is running and integration of modules is going on. Similar strategies in planning, cooperation, coordination, localization and man-machine interaction offer also interesting application potential in industrial, service robotic or educational context. Due to the user interface and data sharing the feeling of being "present" in place for the remote coordinator is increased. This allows a better communication between coordinator and humans in place, as well as an improved coordination of the team.

7. ACKNOWLEDGEMENTS

This work has been supported within the EU-project "PeLoTe – Building *Pre*sence through *Lo*calization for Hybrid *Te*lematic Systems". The authors acknowledge the contributions and discussions with our consortium partners at the Czech Technical University in Prague, Helsinki University of Technology, Certicon a.s. and Steinbeis Transferzentrum ARS.

Fig. 9. Wheeled test vehicle MERLIN without protection cover.

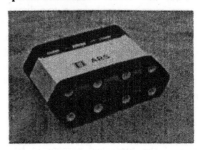

Fig. 10. Tracked MERLIN test vehicle sharing the same electronics and sensor data processing system with the wheeled version, only differing in the locomotion system.

8. REFERENCES

Gilioli M. and K. Schilling. (2003). Autonomous Cooperative Localization of Mobile Robots Based on Ranging Systems. *SPIE Aerosense conference proceedings "Unmanned Ground Vehicle Technology V"*, Orlando, USA, paper 5083-15.

Murphy R., J. Casper, M. Micire and J. Hyams (2000). Assessment of the NIST Standard Test Bed for Urban Search and Rescue. *NIST Workshop on Performance Metrics for Intelligent Systems 2000.*

Saarinen J., R. Mazl, P. Ernest, J. Suomela and L. Preucil. (2004). Sensors and Methods for Human Dead Reckoning. *Proceedings of the 8th Conference on Intelligent Autonomous Systems IAS8, Amsterdam.*

Schilling K. and Q. Meng (2002). The MERLIN vehicles for outdoor applications, *SPIE Aerosense conference proceedings „Unmanned Ground Vehicle Technology IV"*, Orlando, USA.

Wessel I. (2002). *GUI-Design*. Carl Hanser Verlag Munich, Germany. ISBN 3-446-21961-7.

Copyright © IFAC Telematics Applications in Automation and Robotics, Espoo, Finland, 2004

ELSEVIER

IFAC
PUBLICATIONS
www.elsevier.com/locate/ifac

TELEOPERATION WITH A DISTRIBUTED MULTIAGENT-BASED SLAVE MANIPULATOR CONTROL SYSTEM

H.Y.K. Lau and A.K.S. Ng

hyklau@hku.hk, kamseng_alex@graduate.hku.hk
Department of Industrial and Manufacturing Systems Engineering
The University of Hong Kong, Pokfulam Road, Hong Kong, PRC

Abstract: Teleoperation is one of the oldest and presents useful applications in robotics. A teleoperator system consists of a master arm, a slave arm and an operator. Traditionally, slave arm control is achieved by analytical solutions. Multiagent-based paradigm is emerging in Artificial Intelligence. A slave arm can be thought of a group of separately controlled agents that a multiagent-based paradigm can be applied. In this paper, we investigate viability of a multiagent-based slave manipulator control with a cable-driven teleoperator system. *Copyright©2004 IFAC*

Keywords: Distributed Control, Redundant Manipulator, Robot Control, Robotics, Teleoperation

1. INTRODUCTION

A teleoperator system consists of 3 important elements: an operator, a master arm and a slave arm. A master arm is controlled by the operator while tasks are actually performed by the slave arm. Master arm acts as an input device that accepts motion demands from a human operator. The slave arm follows the motion of the master mechanism such that tasks are performed by a slave arm controlled remotely by a human operator through the master arm.

Traditionally, a slave arm of a teleoperator system is controlled by inverse kinematics resolution. (Asada and Slotine, 1986; Craig, 1986; Murray *et al.*, 1994). And in the case of the particular cable-driven teleoperator systems developed in the Intelligence Systems Laboratory in the University of Hong Kong, Lau *et al.* (2002) propose a Jacobian-based control strategy. There are limitations on kinematics solutions. Researchers have to deal with singularities and redundancy. Mori *et al.* (1996) experimented multiagent-based manipulator control with 2D tasks by task decompositions. Ramdane-Cherif *et al.* (2002) proposed multiagent paradigm to solve kinematics solution. However, few studies have been done on applying multiagent-based paradigms to teleoperation. This paper compares the performance of the teleoperator system made with cable-driven

mechanisms at our laboratory. Experimental study based on an implementation of a force reflecting teleoperator system using a closed-form kinematics solution, namely Jacobian transpose, is compared with a multiagent-based control paradigm on the slave manipulator. According to simulation with our MATLAB-based simulator as well as actual experimentation with the control paradigms using the actual cable-driven teleoperator system, the results show that a multiagent-based control has a number of advantages in controlling a redundant slave mechanism in a teleoperator system such as the Whole Arm Manipulator of our system.

This paper presents the teleoperator system that is controlled with a closed-form kinematics solution in Section 2. Section 3 of this paper discusses the performance of the teleoperator system under the control of a closed-form kinematics solution. A Multiagent-based slave control approach is then presented in Section 3 that is followed by the study of deploying such Multiagent-based control approach to control a redundant slave manipulator in Section 4.

2. TELEOPERATION WITH CLOSED-FORM KINEMATICS SOLUTIONS

A typical teleoperator control system derives the slave trajectory based on the command given by the master through well-defined analytical solutions. Inverse kinematics and Jacobian transpose are common approaches to solve such solutions (Craig, 1986; Paul, 1981). The analytic solutions enable computation of essential joint variables for a trajectory including position, velocity and orientation, etc. of a master arm. In particular, Jacobian Transpose maps the Cartesian space velocity to joint space velocity that can be used to approximate the control parameters such as the location of new set points for slave arm control.

2.1 System Configuration

The teleoperator system developed at the Intelligent System Laboratory of the University of Hong Kong consists of cable-driven master and slave arms. The master arm is a SensAble's Phantom joystick (Massie and Salisbury, 1994) that is capable of providing six Degree-of-freedom force-feedback. The slave arm is a four Degree-of-freedom WAM robot (Townsend and Guertin, 1999) which is a fully cable-driven robot capable of whole arm manipulation. In the current setup, the two robot arms are installed in the same room so that the operator observes the task space directly without the aid of monitoring systems. In addition to joint torque sensing, an ATI force torque sensor is installed at the end effector of the slave arm to derive the force feedback signal.

The teleoperator control system is implemented on a real-time QNX operating system with a high bandwidth communication interface developed based on Ethernet. The architecture of the control system is hierarchical, with a number of concurrent tasks running at different priorities. A dedicated joint servo task was created to control each of the joints of the master and slave arms with higher levels of control including trajectory control (using closed-form kinematics solution or multi-agent based solution), teleoperator control and system state control taken care of by higher level interacting tasks. Under normal operating condition, our teleoperation system achieved an outer control loop frequency of 500Hz.

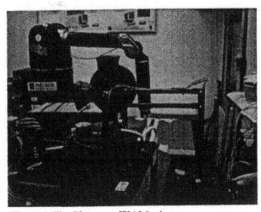

Figure 1 The Phantom-WAM teleoperator system

2.2 Results of bilateral force-reflecting teleoperation

Based in the teleoperator system implemented as defined by Figure 2, the performance of the system under teleoperation control in free space and on hard contact is shown in Figures 3 to 9.

Figures 3 to 6 show the joint-based trajectories of the teleoperator system. It is shown that there exists a small delay between the demand and the actual joint position while the system is in operation. These delays ranges from approximated 86ms to 110ms. Note that joint 3 is intentionally locked as we are not considering redundant control of slave arm in this experiment.

Figures 7, 8 and 9 show the slave Cartesian trajectories and the contact forces measure along the x, y and z-directions respectively based on the ATI force/torque sensor. In this experiment, the slave arm was first moved in free space and the made contact with a wooden block after 9.81 second. The slave arm then slid along the wooden block from 9.81 second to 11.91 second (4905th to 5955th periods) and made another contact from 13 to 14.876 second (6500th to 7438th periods). For each contact, the forces measured are gradually increased from 10 to 30N while the slave arm was moving along the profile of the wooden block. These forces are fed-back to the master arm by the teleoperator system.

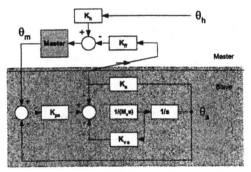

Figure 2 The position-force bilateral teleoperator system

These results show that using a closed-form kinematics solution for slave control achieve satisfactory bilateral force-reflection with a phase lag of approximately equal to 100 ms between the master and slave positions.

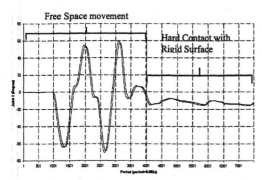

Figure 3 Joint 1 trajectory of slave arm against time (K_{ff}=0.2)

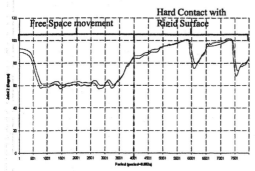

Figure 4 Joint 2 trajectory of slave arm against time (K_{ff}=0.2)

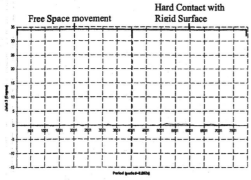

Figure 5 Joint 3 trajectory of slave arm against time (K_{ff}=0.2)

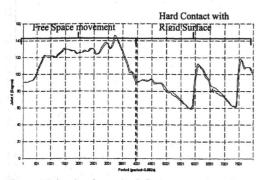

Figure 6 Joint 4 trajectory of slave arm against time (K_{ff}=0.2)

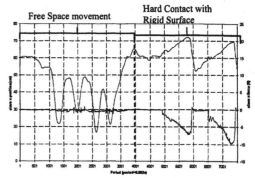

Figure 7 End point trajectory along x-direction of slave arm and force measured along x-direction against time (K_{ff}=0.2)

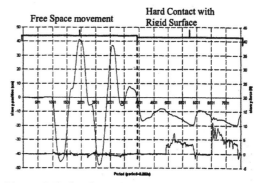

Figure 8 End point trajectory along y-direction of slave arm and force measured along y-direction against time (K_{ff}=0.2)

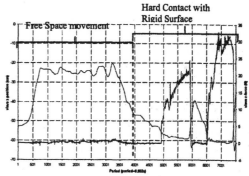

Figure 9 End point trajectory along z-direction of slave arm and force measured along z-direction against time (K_{ff}=0.2)

3. MULTIAGENT-BASED SLAVE MANIPULATOR CONTROL

The multiagent-based control paradigm is a logically distributed system for slave arm control, as each joints are spatially distributed and controlled independently and concurrently. This suggests a totally distributed control paradigm for slave manipulator in our teleoperator system. In a multiagent-based slave manipulator control, paradigm execution of a trajectory involves the interaction between agents. A conventional manipulator is thought of as a system composed of a group of agents, each of which is independently controllable. Conventional methods for robot control have largely been dependent on an exact kinematics solution. In contrast, multiagent based paradigm has multiple agents and each agent may have individual goal(s). All cooperation agents may contribute to achieve a particular goal such as to move the end-effector of a multi-jointed robot along a particular trajectory. In addition, these agents may also aim at improving the overall teleoperation performance.

An agent (e.g. d'Inverno et al., 1998; Dastani et al., 2002; Jennings and Wittig, 1992) can be considered as a logical and autonomous entity which tries to contribute to the overall task achievement based on its own strategy. The agent is assumed to be able to recognize the conditions of the progressing task. Each agent may try to drive the assigned actuators with its own strategy by which it can improve the degree of task achievement. The overall system can be deemed as a set of distributed control variables which are directly or indirectly managed by these agents. Each joint controlling agents decide and carry out their own behaviours based on their intelligence of teleoperator system in response to external sensory signals.

To compare the use of agent-based control with the kinematics control of our 3 Degree-of-freedom teleoperator system as configured in Section 2, a control agent is designed and implemented to each joint. As in the case of using a closed-form kinematics solution for slave control, a 3 Degree-of-freedom agent controller is used in this study. Agents are designed according to the design criteria described in Section 3.1.

3.1 Design of an agent for joint control

1. An agent can only control the movement of single joint variable (θ_i) which cannot exceed joint limit (θ_{si}).
2. An agent only knows about its link length and the control of the joint angle it can vary.
3. An agent can be informed of the locations of the end-effector (x_i) and target (S_i) and the distance ($S_i - x_i$) between them
4. An agent control the variable joint angle according to its objective function:

$\forall \{jt_i\}, i = 1, 2, 3$

$\forall \theta_i(t) \in [-\theta_{si}, \theta_{si}]$

Target, S_i

$S_i = [x \quad y \quad z]^T$

    with reference to local frame $\{i\}$

    s.t. $S_i = S_w T_1^0 \ldots\ldots T_i^{i-1}$

where

    S_w indicates the Set Point in World frame

    T_i^{i-1} indicates the transformation matrix

    derived from Denavit-Hartenberg parameters

Location of the end-effector, x_i

$x_i = [x \quad y \quad z]^T$

    s.t. x_i indicates current position with reference to local frame $\{i\}$

Objective function:

$\theta_i(t) = k_i [S_i(t) - x_i(t-1)]$

where

    $x_i(t) = \text{fwdkinematics}(\theta_1, \theta_2, \theta_3, \theta_4)$

    $k_i = [k_{ix} \quad k_{iy} \quad k_{iz}]$

    s.t. k_i indicates the row vector which is the constant for agent i

3.2 Teleoperation with an agent-based control

From the study, the agent-based control produced similar end-point control as in the case of the closed-form kinematics solution. Figures 10 to 13 show the joint-based trajectory of the teleoperator system, under the multiagent-based slave control. Figure 10 shows the plot of joint 1 trajectory. Further investigation of individual joint-based trajectory shows that the agent-based control paradigm produced a very similar joint trajectory, yet in a more efficient manner in terms of data and computational efficiency. Using a multiagent-based slave manipulator control, joint trajectories are obtained by individual objective functions that are different from the joint trajectories obtained from a closed-form solution. These joint trajectories are 'smoother' indicating that smaller joint travels are required to produced similar end-point movement, hence more efficient control.

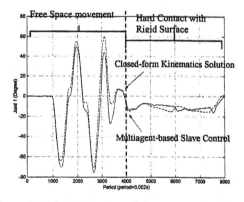

Figure 10 Joint 1 trajectory of slave arm against time

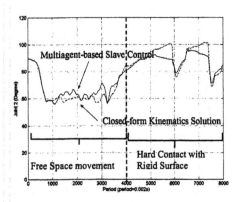

Figure 11 Joint 2 trajectory of slave arm against time

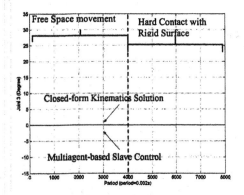

Figure 12 Joint 3 trajectory of slave arm against time

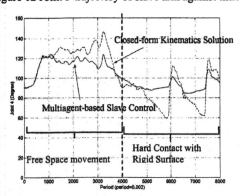

Figure 13 Joint 4 trajectory of slave arm against time

4. MULTIAGENT-BASED SLAVE MANIPULATOR REDUNDANT CONTROL

Section 3 shows that a multiagent-based slave manipulator control gives viable control in teleoperation. The multiagent model is therefore extended to the control of a 4 Degree-of-freedom WAM with redundancy. Each joint controlling agent derives its behaviours according to the agent's objective function that attempt to achieve global objective(s) as defined in Section 3.

Simulation study is performed using a MATLAB simulator developed for the teleoperator system. Using the same set of Phantom (master) arm input, the trajectory demand is fed to the multiagent-based slave manipulator controller to derive the slave trajectory.

Figures 14 to 17 show the simulation results of teleoperation, using the multiagent-based slave manipulator control.

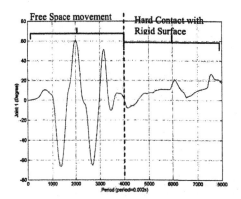

Figure 14 Simulated Joint 1 trajectory of slave arm against time

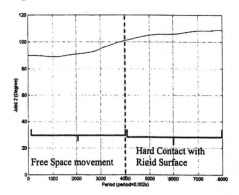

Figure 15 Simulated Joint 2 trajectory of slave arm against time

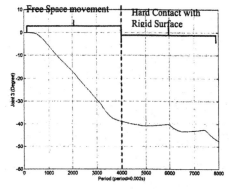

Figure 16 Simulated Joint 3 trajectory of slave arm against time

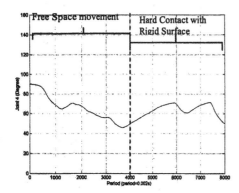

Figure 17 Simulated Joint 4 trajectory of slave arm against time

CONCLUSION

This paper presented an agent-based slave control paradigm that compared with a classical closed-form kinematic approach to bilateral teleoperation. The comparison of teleoperation control under the two approaches namely, closed-form kinematics and multiagent-based slave control is performed. Experimental studies were carried out to establish the practicability of the multiagent approach and it is shown that this approach reduces the computational complexity that does not require computing the pseudoinverse Jacobian matrix. The simulation study shows that desirable joint trajectories were produced. The study was extended to control the redundant WAM and results show that the multiagent approach can control the redundancy without resorting to solving high order polynomial functions. In addition, computational efficiency is improved. These findings indicate that multiagent control provide a promising approach to the control of redundancy and currently, studies to investigate the coordination and cooperation between agents are being undertaken.

ACKNOWLEDGEMENT

The work described in this paper was partly supported by a grant from the British Council and research Grants Council of the Hong Kong Special Administrative Region, PRC, under the UK/HK Joint Research Scheme.

REFERENCES

Asada, H. and Slotine, E. J. (1986) Robot Analysis And Control, 1 ed. *Wiley-Interscience*, 1986

Craig, J.J. (1986) Introduction to robotics- Mechanics & Control. *Addison-Wesley*, 1986.

Dastani M., de Boer F., Dignum F., van der Hoek W., Kroese M and Meyer J.-J.Ch. (2002) Implementing Cognitive Agents in 3APL. *Proceedings of the The 14th Belgian-Dutch Conference on Artificial Intelligence, BNAIC2002*, Leuven, Belgium, 2002.

d'Inverno M., Kinny D., Luck M. and Wooldridge M. (1998) A Formal Specification of dMARS, *In Intelligent Agents IV: Proceedings of the Fourth International Workshop on Agent Theories, Architectures and Languages, Singh, Rao and Wooldridge (eds.), Lecture Notes in Artificial Intelligence*, 1365, 155-176, Springer-Verlag, 1998.

Jennings, N.R. and Wittig, T. (1992) ARCHON: Theory and Practice, *Distributed Artificial Intelligence: Theory and Praxis, pp.179-195, Kluwer Academic Press*, 1992

Lau, H.Y.K., Lau, T. L. and Wai, L.C.C. (2002) A Jacobian-based redundant control strategy. *Proceedings of the Seventh International Conference on Control, Automation, Robotics and Vision, ICARCV 2002*, Singapore December, pp. 1060-1965.

Massie, T. H. and Salisbury, J. K. (1994) The Phantom haptic interface: A device for probing virtual objects. *Proceedings of the ASME International Mechanical Engineering Congress and Exposition*, Chicago, IL, pp.295-302

Mori, A., Naya, F., Osato, N. and Kawaoka, T. (1996) Multiagent-Based Distributed Manipulator Control. *Proceedings of the 1996 IEEE/SICE/RSJ International conference on Multisensor Fusion and Intergration for Intelligent Systems*, pp. 289-296

Murray, R. M., Li, Z. and Sastry, S. S. (1994) A mathematical introduction to robotic manipulation CRC Press, 1994

Paul, R. P. (1981) Robot manipulators: mathematics, programming, and control. *The MIT Press,*1981

Ramdane-Cherif, A., Levy, N., Djenidi, H. and Tadj, C. (2002) Agents Paradigm to Solve a Complex Optimization Problem. *Proceedings of the First IEEE International Conference on Cognitive Informatics, ICCI' 02*, 2002

Townsend, W. T. and Guertin, J. A. (1999) Teleoperator slave – WAM design methodology. *Industrial Robot*, Vol.26, No.3, pp. 167-177

Copyright © IFAC Telematics Applications in Automation and Robotics, Espoo, Finland, 2004

ELSEVIER

IFAC

PUBLICATIONS

www.elsevier.com/locate/ifac

Tele-Experiments Using Satellite Telecommunication Links Based on Multimedia Home Platform Standards

* Klaus Schilling, ° Hendrik Heimer

* Bayerische Julius-Maximilians Universität Würzburg, schi@ieee.org

° Navus GmbH, hendrik.h@navus.de

Abstract: Satellites for multimedia broadcasting are of particular interest to reach in an economical way fixed and mobile users, distributed over large areas. For interactive applications, use of the Multimedia Home Platform standard offers interesting potential. Related hard- and software architecture as well as technology implementation aspects are analysed. At the example of tele-experiments with mobile robots typical application scenarios for remote sensor data acquisition and tele-control are addressed. The application potential of these technologies for industrial telemaintenance tasks is outlined. *Copyright © 2004 IFAC*

Key words: telematics, communication satellites, tele-education, telemaintenance

1. INTRODUCTION

Satellites provide in a most efficient and flexible way the link capacities to interactively connect a far distributed user community without the need of an infrastructure on ground as prerequisite. The ongoing developments in the „Multimedia Home Platform (MHP)"-standard will enable in that context also interactive solutions, offering interesting application potential in provision of services at remote locations.

The specifications of MHP in the mid-nineties included already the potential for encoding transmissions related to transfer alternatives such as satellite, cable or radio broadcasting. Thus a most economical combination of satellite, cable and radio broadcasting for different application scenarios can be implemented. The primary target application area of MHP was the Digital Video Broadcasting (DVB) market. In fact, digital television broadcasting started in Germany already 1996 by using the satellite DF1. Thus the necessary receiver stations had been developed and marketed in combination with the video program offers. Based on that experiences, MHP offers a general standard to extend satellite video broadcasting to a broad spectrum of interactive multimedia services. Therefore the eWave VideoServer (eWVS) project was initiated by the European Space Agency ESA in order to investigate interactive satellite applications, in particular in tele-education.

This paper reviews the technical basis of multimedia satellite techniques and MHP. The options for system architecture and implementation are analysed and finally the application potential will be outlined.

2. SATELLITE BROADCASTING TECHNOLOGY AND MULTIMEDIA HOME PLATFORM STANDARD

Multimedia services (combining different media, such as text, data, audio, video, graphics, pictures in a digital format like MHP) require significantly more bandwidth compared to traditional satellite services such as telephony. Therefore a mix of satellites links with terrestrial high-capacity networks is often considered. Related satellite configurations for this purpose are installed in geostationary and low-Earth orbits (Lutz, Werner, Jahn (2000), Maral, Bousquet (2002)). Typical characteristics of such multimedia satellites are

- use of Ka band,
- multibeam antennas,
- wideband transponders (typically 125 MHz),
- on-board processing and switching,
- large range of service rates (from 16 kb/s to 10 Mb/s),
- low transmission error rates (typically 10^{-10} bit error rate).

Terrestrial data networks are dominated by the Internet Protocol (IP) standard. In order to achieve compatibility with the terrestrial networks and to take advantage in space system development costs, adaptations of IP to telecommunication satellites have been initiated and lead to the Space Communications Protocol Specification (SCPS). Since the 80ies the Consultative Committee for Space Data Systems (CCSDS) developed recommendations for user data packets, packet telemetry, packet telecommand, channel coding and modulation (Lutz, Werner, Jahn (2000)). The data structure is defined by

- *source packets*, including a block of source data, an header with an identifier to route the packet to its destination as well as information about the length, sequence and other characteristics ,
- *transfer frame*, having a fixed length for a given mission providing to the embedded source packets functio-nalities for error control.

Modern satellites have to transfer an increasing amount of multimedia digital data streams (including video and audio) to/from the ground and to other satellites at asynchronous, synchronous and isochronous transmission schemes. Typically MPEG2 coded video streams are transmitted by the Digital Video Broadcasting via Satellite (DVB-S) standard. Further DVB-S with satellite return link (DVB-RCS) has emerged.
Compared to terrestrial networks, a satellite link has to anticipate a signal propagation delay t_D related to the large distances

$$t_D = R / c \quad [s]$$

where R is the range from transmitting to receiving equipment, and c is the speed of light ($3*10^8$ m/s).

Further delays in the different relay links of a telecommunication chain are to be added. This includes delays in terrestrial network, baseband signal processing time, information compression, multiplexing, demodulation, decoding, buffering associated with switching and multiple access, protocol induced delays (in particular for acknowledgement messages). Thus for a geostationary satellite, the minimum signal propagation delay refers to a station in the subsatellite point (where the line connecting the satellite with the Earth's center penetrates the Earth's surface). Here $R_{uplink} = R_{downlink} = 35786$ km. The related overall delay from emission to reception of the signal just related to the distance of the geostationary relay satellite amounts to 238 ms. The maximum delay due to traveling distance occurs, when emitting and receiving Earth stations are located at opposite edges of the satellites coverage area and leads to a delay of 278 ms.

3. MHP VIA SATELLITE

MHP was developed since the mid-nineties as a uniform standard for receiving equipment to provide access to a safe data transfer of a broad range of multimedia applications. To allow only authorized users access to these services, the data are to be scrambled and de-scrambled by a uniform Common Scrambling Algorithm (CSA). Thus for DVB compatible set-top-boxes an uniform Application Programming Interface (API) has been defined and implemented in JAVA (Multimdia Home Platform 1.0.2). A JAVA Virtual Machine serves as basis for software compatibility and for the execution of applications.

The MHP API-specifications refer to 3 application profiles:
- enhanced broadcast profile (without return channel)
- interactive broadcast profile
- internet access profile

The related core functionalities are to be implemented in the set-top-box.

MHP uses packets in specific formats. The application files are transmitted as part of a MPEG-2 stream in a DSM-CC (Digital Storage Media – Command and Control) object carousel. The object carousel is the most common method that is used for broadcasting data in an MHP environment. It provides a broadcast file system which can be accessed by a receiver. The data carousel can be used on its own for getting file data to a receiver. The IP standard is used as basis in the interactive broadcast profile (IP over return channel) and the internet access profile (IP multicast support). Therefore a DVB/IP gateway is to be provided.

Thus for a teleconference application, the MPEG-2-videostreams (embedded into the MHP format) from the different participants, are uplinked to the satellite. It would be most efficient to mix these videostreams in a MHP DSM-CC object carousel, directly implemented on-board the satellite, and to broadcast the result to all receiving stations (cf. Fig. 1). Thus only the delay calculated above would result.

In case the on-board data processing capacity of the satellite is not sufficient, the MHP DSM-CC object carousel is to be implemented on ground (cf. Fig. 2). In consequence the signals have to travel 4 times the distance between Earth surface and the satellite (sender → satellite → carousel → satellite → receiver).

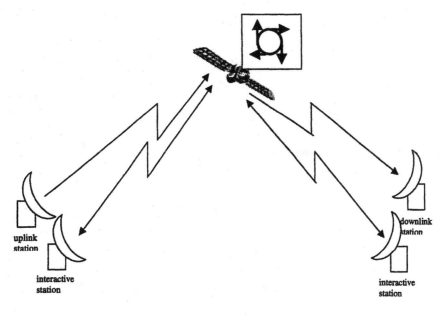

Fig. 1: Configuration with MHP-carousel implemented on-board the satellite: most efficient with respect to delays and bandwidth requirements, but costly.

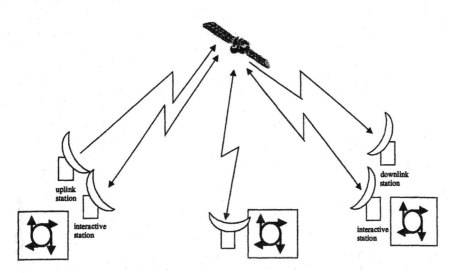

Fig. 2: Configuration with MHP-carousel implemented on each ground station: cost efficient, but more extended bandwidth requirements result.

4. APPLICATION SCENARIOS

The feasibility of the approach based on MHP via satellite is investigated in more detail at the exemplary applications

- tele-experiments in distance education
- industrial tele-maintenance.

In the growing tele-education market a major concern are the high loss rates of students during a course due to missing feedback. Therefore much effort is focussed on increasing interactivity of the students with tutors and with other students. A typical situation is one highly interactive client (the tutor) and several sporadic active clients (the students). The tutor siteoffers here also access to multimedia teaching materials (interactive documents, films, access to experiment hardware,...).

In the demonstrator scenario, students access the mobile robot remote test facility at the University Würzburg (Schilling, Roth, Rösch (2002), Schilling, Meng (2002)) through their TV sets connected to MHP-compliant set-top-boxes. The video signal from Merlin robots are transmitted as a DVB stream with control X-lets intermixed into it. The transmission is initiated by the satellite broadcasting station and is linked through the satellite channel. Figure 3 displays the components constituting the demonstrator scenario which are:

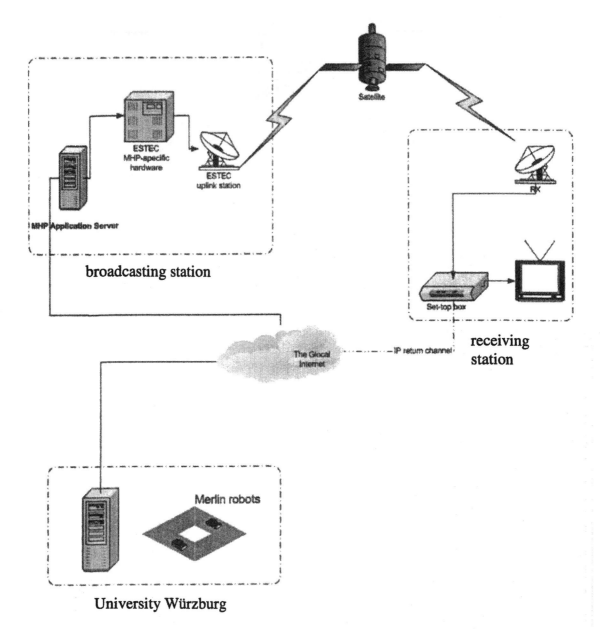

Fig. 3: The tele-education demonstrator to perform remote experiments with mobile robot hardware.

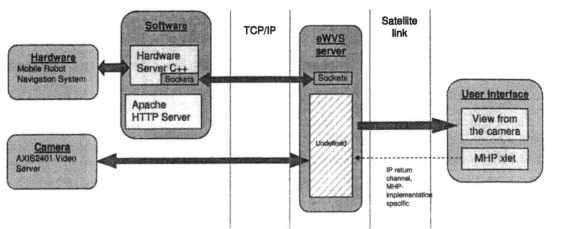

Fig. 4: The software modules to perform the remote robot control experiments on basis of a MHP satellite link by using the eWVS-server.

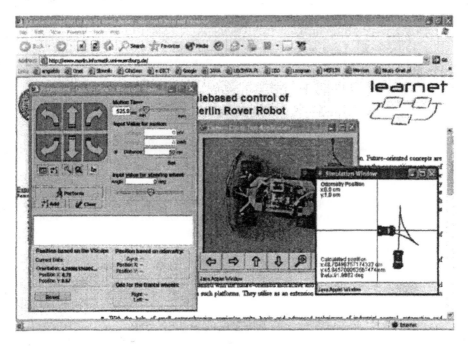

Fig. 5: User Interface to remotely control a mobile robot: at the left the control panel in order to submit commands, in the middle the live webcam image of the experiment, at the right a graphical path planning tool. In addition the user can include JAVA-applets to display sensor data and to access a tutor via chat window.

- the MHP-specific hardware at the broadcasting facilities,
- a co-located PC at broadcasting station running the application server (eWVS),
- a satellite channel used to transfer theDVB signal to the satellite receiver,
- a satellite dish, off-the-shelf set-top-box, and a TV set (to be used at the student sites),
- the mobile robot hardware MERLIN (Mobile Experimental Robot for Locomotion and Intelligent Navigation) located at University Würzburg (Schilling, Meng (2002)).

The same functionalities, related to remote sensor data acquisition and to remote control, form also the basis for industrial telemaintenance applications or emergency support activities. Here the provision of services by remote specialists via fast established satellite communication links is an efficient way to avoid costly stand-still periods in industrial production systems or to provide quick advice and coordination in case of emergencies.

A typical scenario for industrial tele-maintenance is a worker in place, transferring video images and local sensors data according to the advice of the remote specialists. Typically several specialists from different companies and several workers in place observe in parallel the plant equipment performance in order to perform analyses to determine the cause of malfunctions. Usually there are only few workers at the plant location with a highly interactive link demand (for video transmission, transmission of sensor data and receiving voice commands as well as multimedia materials, such as handbooks for repair advice, drawings explaining the equipment design etc.). The remote experts are mainly analyzing the data coming in, and have only sporadic interactivity demands in order to provide advice and information to the workers in place. Thus a similar scenario to tele-education results with respect to the MHP application.

In case of emergencies or catastrophes in a similar way also support for search and rescue activities is to be be provided (cf. Driewer, Baier, Schilling (2004)). As the standard ground based infrastructure might not be available any more, demand for satellite links is obvious in order to coordinate rescue squads from a safe remote site.

5. CONCLUSIONS

The combination of MHP with multimedia-satellite technology provides interesting potential for interactive applications, in particular when

- large areas for fixed and mobile users are to be served economically,
- collecting and broadcasting characteristics are employed,

- networks in short time and in a flexible way are to be established.

A related hard- and software architecture has been developed and applied to a typical tele-experiment scenario, which is also representative for industrial telemaintenance.

ACKNOWLEDGEMENTS

This article is partly based on work elaborated within the project "eWave Video Server", performed in contract for the European Space Agency ESA. The authors acknowledge the contributions and support of the team members A. Fedorov, Y. Khupchenko, Y. Revyakina, P. Douriaguine as well as the ESA officers Nathalie Ricard and Francesco Feliciani.

REFERENCES

Driewer, F., H. Baier, K. Schilling (2004), Robot/Human Interfaces for Rescue Teams, Proceedings *1st IFAC Symposium on Telematics Applications in Automation and Robotics, Helsinki 2004.*

Lutz, E., M. Werner, A. Jahn (2000), *Satellite Systems for Personal and Broadband Communications*, Springer Verlag.

Maral, G., M. Bousquet (2002), *Satellite Communication Systems*, 4th edition, John Wiley & Sons

Multimedia Home Platform 1.0.2 DVB BlueBook A057 Rev. 2

Schilling, K., H. Roth, O. Rösch (2002), Mobile Mini-Robots for Engineering Education, *Global Journal of Engineering Education* 6, p. 79–84.

Schilling, K., Q. Meng, The MERLIN vehicles for outdoor applications, *SPIE conference proceedings „Unmanned Ground Vehicle Technology IV"*, Orlando, p. 43 - 49.

Copyright © IFAC Telematics Applications in Automation
and Robotics, Espoo, Finland, 2004

ELSEVIER

IFAC
PUBLICATIONS
www.elsevier.com/locate/ifac

INDOOR POSITIONING USING WLAN RECEIVED SIGNAL STRENGTH

Jaywon Chey[*], Jae Woong Chun[*], Suk Ja Kim[*], Jin Hyun Lee[**], Gyu-In Jee[**], Jang Gyu Lee[*]

[*] *School of Electrical Engineering and Computer Science, Seoul National University*
[**] *Dept. of Electronics Engineering, Konkuk University*

Abstract: GPS is widely used for outdoor positioning in many applications. However, the signal of GPS is not acquired indoors because of the weakness of GPS signal. Therefore the adequate method for indoor positioning is required. At present, WLAN (Wireless Local Area Network) has been installed in a number of indoor areas such as airport, campus, and park. This paper describes the algorithm using WLAN signal strength to provide the location of the mobile user indoors. Indoor positioning performance is presented by the test experiment and shows the meter-level accuracy regardless of any changes in indoor environment. *Copyright © 2004 IFAC*

Keyword: Positioning systems, Local area networks, Signal levels

1. INTRODUCTION

Recently with the proliferation of mobile terminals such as PDA, cellular phone and so on, various services, for example, indoor navigation for military strategy and fire fighters, commercial advertisement based on indoor location, finding road and so on, are about to be provided via mobile terminal. However, above all, for being provided these convenient and useful services, an indispensable condition is where mobile terminal is accurately indoors. That is, mobile terminal indoor location should be acquired (Abowed, 1998).

The most well known method associated with positioning system is GPS (Global Positioning System). The signal of GPS satellites can be always acquired outdoors and this system provides comparatively accurate location information. However, there are difficulties of applying GPS directly to the indoor positioning because of the weakness of the signal of GPS in indoor environment. And the positioning using the communication signal between base station (BS) and cellular phone doesn't provide adequate accuracy due to the technical limitation of the communication systems and the propagation environments. So, an alternatively new indoor positioning technology to guarantee enough accuracy is required.

WLAN (Wireless Local Area Network) has been installed in broad indoor areas such as airport, campus, and park with interests of mobile internet. Therefore, it is expected that WLAN signals should be easily acquired for the indoor positioning.

The various positioning measurements can be adapted for the indoor positioning like TOA (Time Of Arrival), TDOA (Time Difference Of Arrival) and RSS (Received Signal Strength). First, the characteristic of the indoor environment must be considered to apply these measurements. Severe multipaths by indoor structure exist indoors. TOA measurements need time synchronization between receiver and transmitter and time synchronization on networks should be accomplished in the case of TDOA measurements. But finding the reference time is so hard and, if possible, it is not sure whether mobile terminal acquires at least three APs' (Access points') signal for the positioning under indoor environment or not. Eventually, using TOA and TDOA measurements for the indoor positioning are inadequate because of many problems to solve as mentioned ahead. Therefore, the indoor positioning algorithm using WLAN received signal strength is proposed in this paper.

This paper is organized as follows. In section 2, the related method using WLAN received signal strength and the characteristic of WLAN received signal strength will be explained. The propagation model method is discussed in section 3 and the new positioning method using the propagation model is proposed and the results of the test experiment are shown in section 4. Finally, the conclusions are presented in section 5.

2. RELATED METHOD USING WLAN RECEIVED SIGNAL STRENGTH

The conventional positioning systems with indoor WLAN received signal strength are operated with the database (DB) of the signal strengths that have already been measured at each known sample point. The representative positioning method using the DB is RADAR from Microsoft (Bahl and Padmanabhan, 2000). This method determines the indoor location of mobile user through comparing mobile user's signal strength with signal strength in the DB. Necessarily, the process to create it for indoor positioning is accomplished at the first phase that the system is constructed.

But the WLAN received signal strength is changed temporarily and spatially if considering the indoor propagation characteristics of RF signal. As the measurements collected from one AP for 24 hours are showed in Fig. 1, it seems to be seen that the signal strength is stable. The mean and the standard deviation of the RSS are -41.75 dBm and 1.74 dBm. However, the signal strength often drops suddenly because the indoor environments such as humans, structures, other radio signals and so on, influence the changes of WLAN RSS.

The RF signals can penetrate walls and humans and they cause the increase of the path loss. So the effects of walls and humans have to be considered. The building layout and the construction material make effect on the results in the experiments, too. Human bodies are also obstacles for positioning. The several experiments were conducted and the results of the RSS mean are presented in Table 1 and Table 2. So the orientation of mobile user and other people around mobile user influence the signal strength. The path loss of one human is about 3dBm. These effects should be considered to provide accurate position information.

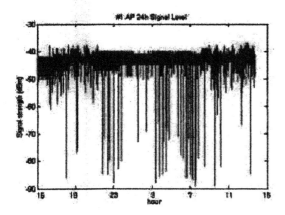

Fig. 1. 24hr signal strength measurements from AP

Like these, because there are many kinds of elements that make effects on the propagation of signal strength, it is impossible that DB is capable of depicting everything anytime and anywhere.

Table 1 Wall effect on signal strength

	Laboratory	Lecture room
Without wall	-41.33 dBm	-39.42 dBm
With wall	-48.71 dBm	-43.87 dBm

Table 2 Human effect on signal strength

	AP #1	AP #2
Without human	-45.25 dBm	-48.10 dBm
With human	-48.71 dBm	-51.39 dBm

3. PROPAGATION MODEL METHOD

The propagation model method is based on the fact that the radio wave loses signal strength as it travels through an environment. The amount of signal strength that the radio wave loses is dependent on the environment. The signal strength is typically modeled by using the log distance path loss model. This model based on both the theory and the measurement, which indicate that the RSS decreases logarithmically with distance, no matter environment indoors or outdoors. The propagation model for the RSS between the AP and the mobile station (MS) of the user is described by equation (1) with the parameter alpha. And the real characteristic of the signal strength over the distance is presented by an experiment in Fig 2.

$$P(r)[dBm] = P(r_0)[dBm] - 10\alpha \log(r/r_0) \quad (1)$$

$P(r)$: signal strength received by a given MS

r: range between AP and MS (m)

r_0: referece range from AP

$P(r_0)$: signal strength at the referece point

α: rate of path loss

The distance from a mobile user to an AP can be calculated by the signal strength loss over space. The triangulation can be used to determine the position of

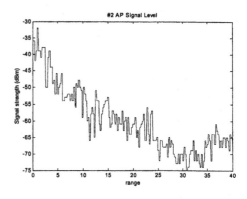

Fig. 2. Signal strength vs. Distance

the mobile user by calculating the distance to three or more APs. This method can determine the position with fairly good accuracy when there is no obstruction and the space is narrow. The accuracy decreases as the distance between the MS and the AP increases. And the accuracy can be improved by increasing the complexity of the propagation model that is used.

However this method has a limitation of the changing RSS over time because of the obstruction, the multipath and so on. Similar to the DB based method mentioned in section 2, the propagation model method stores each sample point's signal strength and computes the estimated parameter in the model equation in advance. Then the location of the mobile user is determined by using the model with the estimated parameter previously and the RSS of the user presently if the mobile user requests his position. The positing accuracy is poor when the present parameter in this model is not same with the estimated parameter beforehand since any modifications of indoor structure change the signal strength received from the sample points. Therefore, regardless of any changes in the indoor environment, another method has to be considered to improve the accuracy.

4. NEW METHOD USING THE PROPAGATION MODEL

To resolve the difficulty of the inconsistent RSS over time due to the changes of the indoor environment, additional hardware is needed. This is the reference station (RS). The RS is installed appropriately in the center of the interested rooms to receive the signal strength well from all APs. The signal strength received from additional RS besides the MS is used to find out the location of mobile user instantly.

This positioning system consists of three parts, which are one MS of the user, several APs and several RSs. It is stored in advance that the known APs' location and the known RSs' location. The RS monitors the signal strength from each APs and the MS monitors similarly at the same time. The parameter is estimated in general propagation model with the signal strength received from the RSs. Next, it determines the location of the mobile user with estimated propagation model and signal strength received from the MS and already known APs' location and RSs' location.

A test bed is established on the sixth floor of the Automation and Systems Research Institute (ASRI) building in the Seoul National University. The layout of this testing area is depicted in Fig 3. It has the dimensions of 17m by 30m with 10 different rooms. The four APs are installed at the locations indicated with star mark and the signal strengths received from the three reference stations are collected at x mark.

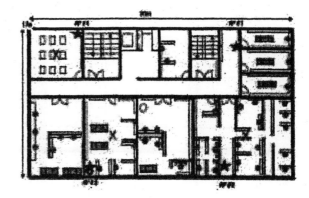

Fig. 3. Layout of the test area

The AP acts as the wireless signal transmitters or the BS and the RS and the MS of the user act as the wireless signal receiver using a laptop computer with Lucent Technology Orinoco wireless card. This network card can detect the signal strength received from the APs. The mobile user is assumed to go around two large rooms, which has the RS. The four points of the user near the each RS is chosen to see the positioning accuracy.

Fig. 4 shows the position accuracy of the proposed method for the indoor positioning in the test experiment. First, the four APs' location is represented as the star mark and the three RSs' location as x mark in this figure. Next, the true position of the mobile user is represented as the circle mark and the estimated position of the user as the square mark. As the estimated position is compared with the true position, the mobile user is located in the same room at least with the meter-level position accuracy. The results of the position accuracy in each test room are depicted in Fig. 5 and Fig. 6.

The position error of the mobile user is shown in Fig. 7. The mean error is 3.184m and the maximum error is 6.045m from the eight test points. The meter-level accuracy for the indoor positioning is induced from this result. This method has the advantages that there is no need to construct the DB of the RSS in many

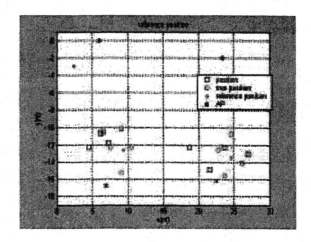

Fig. 4. Results of the test experiment

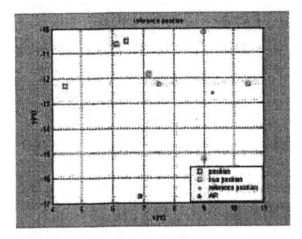

Fig. 5. Results of the first room

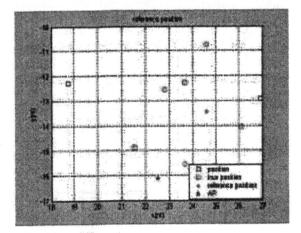

Fig. 6. Results of the second room

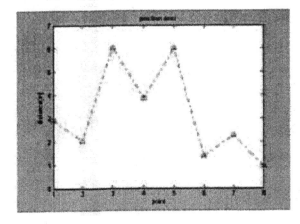

Fig. 7. Position error of the mobile user

sample points in advance and update it periodically. The obstruction such as wall, human and spatial structure in the indoor environment need not be considered for positioning as well.

5. CONCLUSION

The positioning method using WLAN received signal strength is described to provide the position of the mobile user indoors. There are two conventional methods using the WLAN RSS. One method is the DB based method and the other is the propagation model method using signal strength over space. The DB based method need much effort to build the DB of the received signal strength in advance and to update it. The conventional propagation model method does not reflect the instant changes of the indoor structure.

The proposed method using the propagation model uses the additional reference station to reflect the changes of the signal strength that is induced from the modification of the spatial structure over time in the indoor environment. The mobile station of the user and the reference station monitor the signal strength from the APs simultaneously and using the signal strength of the MS and the RSs and known location of the APs and the RSs instantly determines the location of the mobile user.

The accuracy for Indoor positioning is presented by the test experiment and shows the meter-level accuracy regardless of any changes in indoor environment without the burden of building the DB and updating it. Results show the proposed methods can provide a mobile user the reasonably accurate position information.

ACKNOWLEDGEMENT

This work has been supported by the BK21 SNU-KU Research Division for Information Technology, Seoul National University, Automation and Systems Research Institute (ASRI) in Seoul National University and Samsung Electronics Company.

REFERENCES

Abowed, G. D. (1998), Software Design Issues for Ubiquitous Computing, *IEEE WVLSI '98*
Bahl, P. and Padmanabhan, V. N. (2000). RADAR: An In-Building RF-based User Location and Tracking System. *IEEE INFOCOM 2000*

Copyright © IFAC Telematics Applications in Automation
and Robotics, Espoo, Finland, 2004

ELSEVIER

IFAC
PUBLICATIONS
www.elsevier.com/locate/ifac

REMOTE ACCESS TO IMAGES AND CONTROL INFORMATION OF A SUPERVISION SYSTEM THROUGH GPRS

Sempere, V. * Albero, T. * Silvestre, J. *

* Dpto. Communications. UPV
** Dpto Computer Engineering. UPV

Abstract: Real time access to control information of supervision systems in large installations is extremely important in order to be able to act in time to critical situations. In this paper the viability of remote access to this information via GPRS is evaluated. Also evaluated are the request protocol image reception and information about the state of the equipment in the system, as well as the average transfer rate in order to establish the update of the information in real time, obtaining update times, for image and/or control data, which are acceptable for a supervision and control system. *Copyright © 2004 IFAC*

Keywords: Communication networks, Communication control applications, Control system, Images compression, Supervision.

1 INTRODUCTION

In this paper a study is presented on access to control and supervision system information of the purification network of a large city using a PDA through GPRS. The general architecture of the system was presented in a previous article (Sempere, *et al.,* 2003), where the improvements achieved were described, since initially "polling" (Ibe and Trivedi, 1990; Haverkort, 1990) through radio frequency was used for communication between the central and remote stations, presenting poor reliability and low speeds. Currently, through ISDN, the system can exchange control information concurrently, from multiple stations, directly from the sensors and from the installation, or point to point from whatever control equipment (PLC, terminal, etc). It can also capture images of critical zones within the installation and transmit them to the central station where they are processed.

The paper is structured in the following way: In section 2 can be found a general description of GPRS technology. In section 3 the simulation environment and the protocol for image request from the PDA is described briefly. Some measurements are shown in section 4, and in section 5 the conclusions obtained are presented.

2 GENERAL PACKET RADIO SERVICE (GPRS)

GPRS is an extension of GSM mobile communication technology. GPRS has been designed for applications that go further than just voice, known as Advanced Services in Information Mobility.

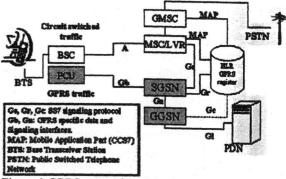

Figure 1 GPRS network architecture

GPRS uses the existing GSM network and adds two new packet-switching network equipment: GGSN (Gateway GPRS Support Node) and SGSN (Serving GPRS Support Node). In figure 1, it can be seen that GPRS introduces at the BSC level (Base Station Controller) the unit known as PCU (Packet Control Unit).

GGSN is similar to the GMSC (Gateway Mobile Services Switching Center), it acts as a logical interface toward the external PDN (Public Packet Data Network) or other GPRS networks. The GGSN connects with the HLR (Home Location Register) by means of the Gc interface. The SGSN controls the connection between the network and the mobile station (MS), and is in charge of the delivery of packages to the mobile terminal in a way similar to that of the MSC (Mobile Switching Center) and the VLR (Visitor Location Register) made in the GSM. The following are characteristics of the GPRS:

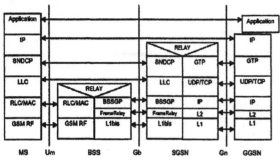

Figure 2 GPRS protocols

- It is an always-on service. Users only pay for the amount of data they transfer rather than for the length of time they are connected to the network.
- It meets the Internet communications protocol (IP).
- It uses advanced codification with different classes and greater speed than GSM.
- It allows linking of time intervals to increase capacity.
- The GPRS system is completely compatible with voice transmission via GSM.

2.1 The GPRS Protocols Layers

Figure 2 shows each one of the protocols that intervene in the transmission of data and signalling information in GPRS. Following this, there is a brief description of each protocol. [Agilent Technologies, 2001; Ericsson, 2002, Ghribi and Logrippo, 2000; Granbohm and Wiklund, 1999]:

- GTP (GPRS Tunnelling Protocol): Receives IP datagrams and sends them to the GPRS support nodes.
- TCP/UDP (Transmission Control Protocol and User Datagram Protocol): TCP is used to transfer Protocol Data Units (PDUs) with reliability and UDP is used to carry the information that do not require reliability.
- IP (Internet Protocol): This is in charge of routing user data and signalling information across the Gn interface.
- SNDCP (SubNetwork Dependent Convergence Protocol): This protocol is used to convert the network layer PDUs (N-PDUs) into an appropriate format for the GPRS architecture.
- LLC (Logical Link Control): This protocol obtains a reliable logical link between the SGSN and the mobile phone.
- BSSGP (Base Station System GPRS Protocol): Sends information between the SGSN and the BSS.
- NS (Network Service): Uses frame relay across the Gb interface. It could be used like a point to point connection between the SGSN and the BSS or as a frame relay network.
- RLC (Radio Link Control): Transfers LLC-PDUs between the layer an the MAC function, segments the LLC-PDUs into RLC data blocks, segments and regroups the

RLC/MAC messages into RLC/MAC control blocks.
- MAC (Medium Access Control): This protocol controls the access signalling across the air interface.

2.2 Data transfer

Having defined the different GPRS protocol layers, how packet data units (PDUs) are transmitted (Fabri, et al., 2000; 3GPP, 2002; Inacon, 2000) is going to be described. The data has to travel from the Internet to the GGSN, then to the SGSN, to the BSS, and finally to the MS (in this paper the PDA). Along the way the data suffers a series of changes, i.e., headers are added in each of the layers, also occurring are segmentation and even compression if necessary.

In figure 3 data flow is shown and it is possible to appreciate that is rather header-heavy, therefore the available physical link layer throughput is occupied by header bits, rather than information data.

The SNDCP is responsible for a number of functions: compression and decompression of user data, compression and decompression of packet headers (this function is only applicable in case of TCP/IP), segmentation and desegmentation of N-PDUs, (Inacon, 2000).

The LLC layer concatenates the radio blocks to/from the RLC/MAC layer into larger segments (LLC frames). A LLC frame has a variable length and consists of a header (address (1 octet) + control field (1-36 octets)), an information field with the user data to/from the upper layers (140-1520 octets), and a Frame Check Sequence field (FCS, 3 octets), (3GPP, 2002).

The RLC/MAC data is transmitted in radio blocks that use a structure with a MAC header. There are two types of blocks: the control blocks, with signalling information and the data blocks, with user data. The size of the radio blocks depends on the Coding Scheme (CS) used in the transmission, a control block is always coded with CS-1 while data can be coded using any of the four coding schemes, (Ericsson, 2002). In this paper the CS-2 which uses the PCMCIA GSM/GPRS card installed in the PDA is going to be considered. The RLC/MAC block is composed by the header block (MAC header[1] 1 octet, RLC header 2 octets), some spare bits added (7 bits) it is a padding needed to adapt the radio block size to the channel coder, the RLC payload data block (30 octets), the Block Check Sequence (BCS, 16 bits) and after all a tail bits (4 bits). Each radio block is assembled differently depending on the coding scheme used, however, regardless of coding scheme, each radio block must be made 456 bits long in order to fit it into the GSM bursts.

---

[1] *In the byte which forms the MAC header 3 bits correspond to the USF (Uplink State Flag), this is precoded with a block code in the CS-2, CS-3 and CS-4 in order to detect any errors in it. Therefore the precoded USF is 6 bits and the MAC header for the CS-2 is (8-3)+6=11*

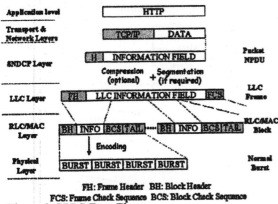

FH: Frame Header   BH: Block Header
FCS: Frame Check Sequence   BCS: Block Check Sequence

Figure 3 GPRS Data Flow

3   EXPERIMENTATION ENVIRONMENT.

The system in which the tests have been carried out is made up of 3 fundamental parts, see figure 4. For a detailed description of the complete system see Sempere, *et al., 2003*:

- Central station.
- Remote stations.
- Communication network.

3.1   *Central station*

This is the part charged with managing the information coming from all the installation and controlling the remote communication between the central station and the remotes. It communicates simultaneously (using several connections TCP/IP) with all the remote stations.

The central station uses equipment which does the SCADA (Luque, *et al., 1996*) in the new network and houses the state and image data base. A background process is charged with preserving coherence between the old data base (updated by polling) and the new one.

Local application of the central station. In the application of the central station is found a map of the city where the different stations are located, when being placed on a station a pull-down menu appears with the options to execute.

The parameterization of the variables to read (states) and to write (orders) from the PLC is made in the central station. In this configuration, the position in the zone of memory or marks that will contain the order/state information is indicated. The initial values of the states zone are stored in the data base when finishing the parameterization, and later each change that is detected is stored. Another option is to visualize the states of the equipment of a remote station and give orders from the central station.

The images are visualized in the central station in real time connecting itself to a remote station, if the operator considers it necessary he can store them. Another type of storage is that made in background. The parameterization of the images is made and the recording is made with a constant regularity

throughout the day. Later the stored images can be recovered (background or manual).

Web application of the central station The central station also houses a Web server which permits the data base which registers information from each remote to be consulted by Internet.

Access to the information can be made from fixed or mobile nodes using GPRS. For this a Web page has been designed which in the moment of connection detects and adapts to the type of user whether it is fixed (PC) or mobile (PDA). The application has been developed with JavaScript, VBScript, ASP and HTML. For PDA JavaScript 1.0 is used due to superior versions not being accepted.

The application Web security is one of the important aspects. For this there is a system of authorized users which restricts open access. Moreover a Web user can only consult information which has no control over the system.

If the client connects from PC, he is permitted to see control data, images in real time, and stored images. The client can gain access to this information in two ways, either by way of a tree-control, or through the interactive map that calls up a menu. The real time images are updated in a fixed time and it is possible for the client to capture an image that interests him and store it. The control data option has the task of showing the state of the equipment that makes up each station.

If accessed through a PDA client, the images can be seen live, as well as the information on conditions in real time, Fig. 5. PDA, Compaq iPAQ 3970 has been used.

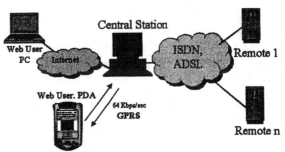

Figure 4 Scheme of the simulation environment

Figure 5 Web Application for PDA client. a) visualization of control data, b) visualization of images in real time.

The decision to create a Web application is determined by the elevated price of a point to point GPRS connection with the central station or with the remotes through which the mobile users access information. For this reason, it has been decided to use the Internet as a transport network and GPRS as access technology.

3.2 Remote station

It can capture information proceeding from different cameras situated in the installation, then process and transmit it together to the central station by means of the communication available in each case. It is capable of communicating bi-directionally, with any PLC on the market (the Simatic S5/S7 protocol has been used in the prototype) and mapping the variable principles of the process in a simple way. The process of the remote station determines when variations have been produced that are susceptible to being sent to the central, carrying out an automatic update.

Application of the remote station. There are four process in the remote station: *"CapturaImag"*, *"ClienteImag"*, *"ComprImag"* and *"ControlCam"*. The first is the task of capturing the images from the camera and sending them to *"ComprImag"*, which compresses them. *"ClienteImag"* transmits them to the central as well as receiving parameterisation information and the camera movements requests from the central station. These requests are transmitted to *"ControlCam"* which schedules the camera. The four processes share a common memory zone that allows them to communicate in a synchronised manner. The central carries out the parameterisation and indicates when the transmission/reception can be initiated or delayed, using messages to the *"ClienteImag"* process.

The communication of the remote with the PLC is carried out using the PlcS5 ActiveX Control, which uses the Siemens Simatic S5 AS511 protocol over a series link. In the case of a change taking place in the PLC, an alarm signal is sent from the remote station to the central station.

3.3 Communications network

This gives communication support to all the remote stations, permitting the incorporation of services in real time such as the transport of images, states and orders. The previous system of polling using radio frequency has become a redundant back up system, which operates in the case of failure with minimum service.

Three alternatives have been studied for the interconnection of the central station with the remotes: a) Use of point to point connections through the RDSI (Integrated Services Digital Network), b) VPN (Virtual Private Network) (Kosiur, 1998) through the Internet with RDSI access and c) VPN through the Internet with ADSL access. GPRS has been used for the interconnection between the central station and a mobile Web user (Hoymann and Stuckmann, 2002).

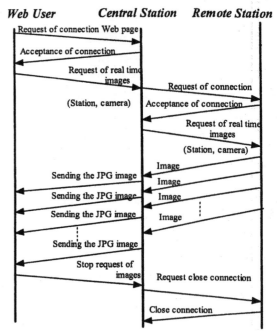

Figure 6 Protocol for the request and reception for a Web user.

3.4 Protocol for the request and reception of images and control information via GPRS

Once the environment in which the tests have been carried out have been generally described, the protocol for the request and reception of images is described, figure 6, where the finished study is centred. This same protocol is similar to the one carried out in the request of a Web user to visualise the PLC states.

When a Web user wants to seek information on a remote, whether on PLC states or images, he must be connected to the Web page and be identified. Once the access to the Web page has been permitted, the request can be made. This request is picked up by the local program of the central station through a message sent by the Web application.

The central station then requests connection to the remote station so as to offer the information in real time to the Web user. When the connection is established, the remote station sends the requested information to the local application of the central. In the case of states, if there are changes, they are stored in the database, and in the case of images, these are compressed creating JPG files, which use the PDA to visualise the requested images.

4 EXPERIMENTS AND RESULTS

The characteristics of the PCMCIA GSM/GPRS Xacom card used are the following:
- o Class 12 SW (4+4, 5 simultaneous).
- o Transference velocity 67.0 Kpbs in Coding Scheme 2.

The measurements of the transference rate were made using two different methods, obtaining very similar results in both. The first method used was the connection using GPRS to a Web page that measures the transference speed (Kbps). The steps to carry out the measurements are the following: A primary estimation of the speed is made sending a page of a determined size. Measuring the time taken to download the page, a primary result of the speed of download is obtained. Secondly, the real measurement is carried out by sending a page of an appropriate size at the speed calculated in the previous stage. In the second method an FTP server is used. A file of fixed size is downloaded from the FTP client of the PDA, and the time and size of the download is obtained, calculating the rate in Kbps.

Measurements were also taken of the transference rate on different days. In figure 7 the variability of the transference rate (Kbps) over a period of 5 days is shown, where the measurements have been taken between 9:00 hrs and 18:00 hrs with intervals of 15 minutes in the city of Alcoy. Also, the deviation from the average with respect to the maximum and the minimum of each instant of measurement is rather high. Measurements were also taken to obtain the transference rate in three other cities and in different time zones (13:00 – 20:00, 20:40 – 23:45, 7:30 – 9:00), figure 8.

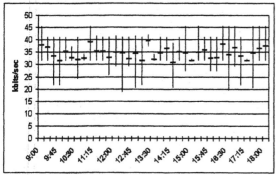

Figure 7 Graph which shows the average, maximum and minimum value, over a period of five days for each instant of measurement, between 9:00 and 18:00.

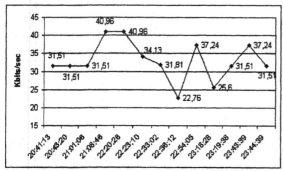

Figure 8 Transference rate obtained with measurements taken between 20:40 and 23:45 for a single day.

Table 1 Average value of the transference rate (Kbit/s)

City	Day1	Day2	Day3	Day4	Day5
Alcoy	33.5	30.67	36.07	35.73	36.19
Valencia	34.69	34.30	32.41	36.06	35.35
Banyeres	35.87	32.94	38.21	36.48	35.89
Játiva	32.62	36.79	35.90	31.72	33.52

According to figures 7 and 8, and table 1 where the average speed obtained from the different measurements is shown in the area where the tests have been carried out, the average speed which is indicated in the characteristics of the card is not reached. As is known, the transference rate as well as the characteristics of the card depend on the following: GSM users who are using voice, the GPRS users who are connected at that moment, the geographical area, and last but not least, the operator who offers the service, in the case of the tests, Telefónica Movistar.

Bearing in mind that the JPG image that is visualised on the PDA Web page has a size that varies between 4 and 7 Kbytes, the time to download the image, taking an average rate of 34.46 Kbits/sec, varies between 0.91s and 1.63s (Tu, Unloading time). Also, an Internal Processing Time (TIP) exists, in which the defragmentation and decompression tasks are included for the visualisation of the image on the PDA. This can last between 1 and 1.5s. Another time to bear in mind is that which the user needs to be able to observe the image (Tv, Visualisation Time); which various observers believe should not be inferior to 3s.

Another aspect to consider is the following: due to the special characteristics of wireless links when TCP protocol is used in GPRS networks, it may not be able to achieve expect performance (Gurtov, *et al.*, 2002). The wireless links are prone in general to a higher transmission error rate (interference, fading and poor signal quality, packet loss contributed to node mobility). TCP incorrectly interprets packet loss or delay as the results of congestion, then TCP proceeds to congestion control and the throughput is affected. For this reason, a propagation delay (ΔT) is considered between 0.066 and 200 ms (Zhou, et al., 2002).

Considering the worst case, the time (T) that must be established for the updating of the image on the PDA web page follows the expression below:

$$T = T_U + T_{IP} + T_V + \Delta T$$
$$T = 1.63 + 1.5 + 3 + 0.2 = 6.33 \text{ s}$$

If this same time is calculated for the best case, the result obtained is 4.91s. However, different tests have been carried out with this time, and in many of them it has not been sufficient for the image to be shown on the screen. With a time of 6s the images are visible and the observer can analyse them. Therefore, this updated time has been finally established for the image shown on the PDA.

5 CONCLUSIONS

A viable system has been presented to which one can gain access via GPRS to obtain control information (states of equipment and images of the remote stations in real time) at any time of day.

As regards the images in real time, the visualisation time Tv, which the user has available to observe the image, can be inferior or superior to the time considered necessary (3s). This is due to the fact that the variation in other times for various reasons (congestion, transference speed, errors in transmission, etc.) has repercussions on the value of Tv. The user, therefore, sometimes does not have the necessary time available to observe the image on the PDA completely.

The total time necessary for the update of information about states in real time is 4s, which is inferior to that necessary to see images in real time. This is due to the fact that the information about a change in state contains little data to transmit, so the reception of the said values is faster, the possibility of error is smaller and also the time necessary for observation on screen is shorter.

It has been confirmed that the Telefónica Movistar operator offers an average rate of 34.46 Kbits/sec and this never goes above 40.96 Kbits/sec.

In later studies, an analysis of the behaviour of the system is intended, using another GPRS operator such as Amena or Vodafone. After analysis, the operator that offers the greatest transference rate will be chosen, to shorten the refreshment time established for both images and states.

In future studies it is also intended to provide greater security for the PDA and PC Web application, so as to have control over the system. That is, to be able to give orders to the remote (open/close sluices, start/stop pumps, etc) and to control cameras (zoom control, brightness and backlight and camera movement), and not to only obtain information as happens now.

This will be possible since the Web server used is Internet Information Services (IIS) 5.0 on Windows 2000 Server which provides security in the communications. In the first place it has an encrypted authentication system that already is being used in the Web application, and on the other hand allows the encryption of the data that is transmitted during the communication by means of SSL 3.0 (Secure Socket Layer) and TLS (Transport Layer Security), which needs a Digital Certificate. In addition the server will be configured to grant or to deny access to the Web according to the IP address of the user, allowing only connection from authorized computers.

6 ACKNOWLEDGEMENTS

This work was supported by the "Ministerio de Ciencia y Tecnología" of Spain under the project TIC2003-08129-C01, which is partially funded by FEDER and by the "Generalitat Valenciana" with the grant TS/03/UPV/09.

7 REFERENCES

Agilent Technologies (2001). Understanding General Packet Radio Service (GPRS). Application Note 1377. http://agilent.com

Ericsson (2002). GPRS Measurements in TEMS Products. White paper http://www.ericsson.com/services/tems/whitepapers.shtml

Fabri, S.N., S. Worrall, A. Sadka and A. Kondoz. (2000). Real-Time Video Communications over GPRS. *3G Mobile Communication Technologies, Conference Publication* No. 471. pp. 426-430.

Ghribi, B. and L. Logrippo (2000). Understanding GPRS: the GSM packet radio service. *Computer Networks* 34, pp. 763-779.

Granbohm, H., J. Wiklund (1999). GPRS General Packet Radio Service. *Ericsson Review*. No 2.

Gurtov, A., M. Passoja, O. Aalto and M. Raitola (2002). Multi-layer Protocol Tracing in a GPRS Network. *IEEE Fall VTC*.

Haverkort B. R. (1999) Performance Evaluation of Polling Based Communication Systems Using SPNs. In: *Application of Petri Nets to Communication Networks: Advances in Petri Nets* (J. Billington, M. Diaz and G. Rozenberg, Eds) 1st ed. pp. 176-209. Springer. Berlin.

Hoymann, C. and Stuckmann, P. (2002) On the Feasibility of Video Streaming Applications over GPRS/EGPRS. IEEE *Global Telecommunications Conference*.

Ibe, O.C. and K.S. Trivedi (1990). Stochastic Petri Net Models of Polling Systems. *IEEE Journal on Selected areas in Communications*, Vol 8, No 9, pp. 1649-1657.

Inacon. (2000). GPRS From A-Z. http://www.inacon.com

Kosiur, D. (1998) *Building and Managing Virtual Private Network*. Wiley.

Luque, J., J.I Escudero. and I Gómez. (1996). Determining the Channel Capacity in SCADA Systems Using Polling Protocols. *IEEE Transactions on Power Systems*. Vol. 11, No. 2, pp. 917-922.

Sempere, V., T. Albero and J. Silvestre (2003). Supervision and Control System of metropolitan scope based on Public Communication Networks. 5th IFAC Conference on Field bus Systems and their Applications. pp 317-323.

Sempere, V., T. Albero and J. Silvestre (2004). Analysis of Communication alternatives over Public Networks for a Supervision and Control System of Metropolitan Scope. IEEE International Symposium on Industrial Electronics. May 2004.

Zhou, L., P. Chan and R. Radhakrishna (2002). Effect of TCP/LLC protocol interaction in GPRS networks. *Computer Communications* 25, pp 501-506.

3GPP (2002) Digital Cellular Telecommunication System (phase 2+); Overall description over the GPRS radio interface; Stage 2; (3GPP TS 43.064 version 5.0.0 Release 5).

Copyright © IFAC Telematics Applications in Automation
and Robotics, Espoo, Finland, 2004

WEB BASED REMOTE CONTROL OF
MECHANICAL SYSTEMS

Philippe Le Parc, Jean Vareille, Lionel Marcé

Université de Bretagne Occidentale
EA2215 - Langages et Interfaces pour Machines
Intelligentes
20 av. Victor Le Gorgeu, BP 809
29285 Brest Cedex - France
Philippe.Le-Parc@univ-brest.fr

Abstract: The control of mechanical systems is nowadays local. In the future, with
the development of the new e-technologies, this control will be remote and will
be made over networks, even unpredictable like Internet. In a first part of this
paper, we are presenting the main interests of such a control and we are describing
some existing applications. In a second part, we are proposing a generic software
architecture to realize such remote control and a methodology to take into account
the unpredictable nature of the communication media. *Copyright © 2004 IFAC*

Keywords: Remote Control, Internet, Methodology, Software Specification,
Networks.

1. INTRODUCTION

Since 1970s, the development of the Internet is
certainly one of the major changes of the industrial era. The Internet network is now:

- Accessible from nearly every point of the
  Earth, using various technologies from telephone lines to satellite transmission systems.
- Becoming faster and faster, cheaper and
  cheaper, day after day, with the developing of
  new technologies and the increase in number
  of Internet links.
- Easy to use and also easy to deploy: navigators are now belonging to classical software
  shipped with any computers and web servers
  are also following the same way.

As scientists, we do not only have to follow these
changes but to offer new uses of this technology.
Offering new uses, is not only putting the "Internet" word everywhere and developing some basic

web-servers just to provide information, but is
integrating within our fields of competence the
changes implied by Internet.

In the field of automatisms, the emergence of
Internet is shown by the use of the IP technology,
which starts to replace or to be an alternate
solution to industrial buses like Telway, Profibus
or DeviceNet. This change is governed by the fact
that IP can be considered now as a standard
protocol and that a large amount of tools have
been developed around it. One direct consequence
is that it becomes really easy to install micro web
servers (or embedded web servers), as example,
to propose some html pages to look at data
coming from various sensors in the way to follow
production processes.

The next industrial step will be not only to get
information but also to send information to the
production line using IP links. And the next step,
will be to make all theses operations, not only using IP links but really being on the Internet, that

is to say, to be able to control a production process from any point of the world using usual tools like a laptop and a telephone line. But, the problem that will still remain, is the confidence user can have in the Internet in terms of Quality of Services (QoS). Right now, no quality is guaranteed and in the future, quality could be guaranteed but it will be expensive. Without this QoS, it becomes really hazardous to perform remote control of industrial processes.

Nevertheless, in some fields, using Internet technologies to control production lines or mechanical systems [1], is possible, for example in the following fields:

- Tele-teaching: a lot of universities are using machines to teach the basics of mechanical engineering. The profitability of these machines is of course really poor because they are only used a few weeks a year. Why not developing common centers, where students may have access to real machines without being close to them ?
  One of the problems of e-learning is to make practical experiments. Why not using Internet technologies to let distant students to manipulate real systems ?
- Tele-maintenance (Bicchi *et al.*, 2001): with Internet technologies, it is now possible to make remote diagnostics, to solve and repair problems, to prepare maintenance phases etc...
- Tele-expertise: some specific operations on mechanical systems can only be made by expert. In a close future, it will become possible for experts to operate from their office a machine located somewhere in the world, just using classic web technologies.

The use of Internet will reduce the costs of these activities. The increase of Internet abilities in term of speed and bandwidth in the future, let us also think that the quality of the remote control and the comfort of the user will also increase. But, when developing such applications, we have to think that these activities rely all the time on an unpredictable network and that we have to build them taking into account this parameter.

In this article, we will first describe some of the major experiences about systems of tele-control, which have been made. Then we will propose a method to take into account quality changes in the Internet network when controlling machines and a generic reusable software architecture for this purpose. To conclude, we will present some applications and also some future developments

---

[1] we will use in the term "machines" to describe both production line and mechanical systems in the rest of this article

2. SOME INTERNET EXPERIMENTS

In the 90's, some projects of remote control of mechanical systems, using Internet as communication network, emerged: the Mercury project (Goldberg *et al.*, 2000), the Australian Telerobot (Taylor and Dalton, 1997), KhepOnTheWeb (Saucy and F.Mondada, 2000), Rhino (Burgard *et al.*, 1995), Xavier (Simmons, 1998), Ariti (Otmane *et al.*, 2000), Puma-Paint(Stein, 1998) etc...

They more or less have the same structure: the controlled mechanical system is connected to a computer, which runs a piece of software acting like a server. This one may execute requests made by a distant user who is basically running an Internet navigator. On the client side, CGI scripts, used at the beginning, have been replaced mainly by Java applets to propose more sophisticated Human Machine Interface and also to work in a "connected" manner. Underlying network technologies used are based on IP: generally associated with TCP for commands and with UDP for flow of information (like video).

The different authors of these platforms have focussed their research on some specific topics such as proving it was possible (Mercury), user interaction and documentation (Telerobot), mobile robotics (KhepOnTheWeb, Rhino) or augmented reality (Ariti).

All these experiences are really interesting because they have treated the problem of remote control in different manners and in different contexts. To develop safe and evaluated web-based remote control, one has to take into account all of them. Nevertheless, we can notice that none of these works have tried to develop some generic software architecture and the unpredictable nature of the network (Internet) is not really taking into account with all the consequences. In the next sections, we will propose some solutions to these two problems.

3. NETWORK QUALITY

The Internet network is working using the "best effort" strategy, which means that any user may send information and that there is no priorities between them. This strategy implies that characteristics for a connection between a user (or client) and a server is depending on, of course, some physical constraints (mainly type of the network, distance, number of nodes) and also depending on the overall use of the Internet. It is then really difficult to propose a model for the Internet network and in many cases, Internet has to be considered as a black box, which takes inputs and may deliver outputs after some time.

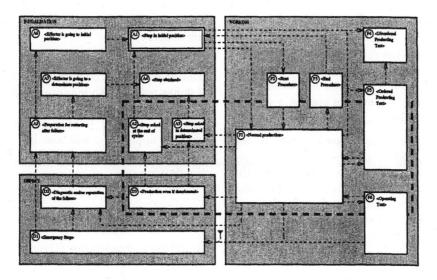

Fig. 1. Gemma description sheet

Nevertheless, two results have to be taken into account:

- The IP communication model uses two protocols : UDP and TCP. The first one is used to send information without any control: packets are sent and may arrive to their destination. More the ordering is not guaranteed. The second one guaranties that packets will arrive and will be delivered to the users in the right order. Moreover, if the connection fails, the sender will be informed (Benali *et al.*, 2001). The communication first starts with a negotiation between the sender and the receiver in order to try to have a regular transmission (in terms of packet per time unit).
- Statistical studies and tests, show that, using TCP, the time to send a packet is rather stable: if a packet needs x ms, the next packet will nearly need the same amount of time (Oboe and Fiorini, 1997).

In the context of remote control, the choice of TCP to send orders is obvious to be sure that orders are transmitted as well as to have a rather stable transmission delay. A way to measure it, is to use a Ping/Pong method: one side of the system send a request (*ping*) and the other has to answer it immediately (*pong*). This method is easy to implement, does not disturb the general working of the system (client/server/network) and gives information, which may be exploited in two different ways:

- Statistical way: when the *pong* is coming back, the measure made corresponds to the previous state of the connection. This information, associated with previous ones, gives to the server and the client a general idea of the quality of the connection which may have an impact on the remote control and the way

a user will be confident in the system: for example, he will not try any critical operation if its connection looks bad.
- Dynamical way: when one side of the system is sending a *ping* (or *pong*), watchdogs may be setup to monitor the waiting time. If it overruns some predefined limits, some autonomous decisions may be taken to prevent any damage on the whole system.

In the following, we will describe a methodological tool, Gemma-Q, that may be used to specify how to take into account network distortion in a remote control application.

4. MANAGING NETWORK DISTORSION

As shown in the previously, the Ping/Pong method acts as a sensor, which gives information about the network quality. Basically, one can decide that the connection is correct when the delay observed is lower than a limit and incorrect in the other cases. This implies two different states for the system under control and transitions between them.

More generally, mechanical systems may have different states. A specification tool called Gemma [2] has been defined to model these different states. It is composed of three main areas (Figure 1): Initialization, Working and Defect. Each area is itself composed of sub-areas, which corresponds to sub-cases. For example, the Working area is composed of 6 sub-areas like "F1: normal production", "F2: start procedure" or "F6: operating test". When an engineer has to specify the way a mechanical system is working, he will describe, in each interesting sub-area, what the system will do

---

[2] Gemma is a French acronym for "Guide d'Etude des Modes de Marches et d'Arrêt"

and also the transition condition (boolean expression depending on input or interval values) that may exist between the different states. Moreover, the "D1: emergency stop", is a specialized sub-area that may be reached from any other sub-area. The Gemma tool may be seen as a generic Statechart(Harel, 1987).

A first idea, to add the network aspect to the Gemma model, will be to create a specific sub-area, like the D1, that may be reached when the network will be recognized as unusable. This idea does not take into account the fact that we can define several (not only two) quality levels.

A new generic Gemma(Ogor, 2001) has been defined: each sub-area is splitted in sub-sub-areas, connected by specific transitions, depending on the network quality (Figure 2).

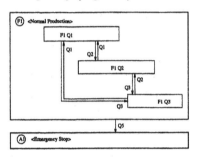

Fig. 2. F1: normal production : Gemma-Q model

Six different quality levels have been defined:

- **Q1:** corresponds to a very good quality, which means that a user may really have a good control on the system.
- **Q2:** corresponds to a rather good quality. The user may encounter problems if the quality decreases.
- **Q3:** in this state, the quality is considered as not really good for the people in charge of the system. But, one may imagine that someone else, with a better quality, may take the control of the system. A search is then started, to find the best controller.
- **Q4:** corresponds to a very poor quality, but the user may still send some orders. He is, of course, supposed to send "stop" command.
- **Q5:** in this level, the connection is considered as broken. Automatic actions have to be done to put the tool in a safe situation and to inform the user.
- **Qz:** this level has been defined to take into account the possibility that a system work in an autonomous way, without any user in charge of it. It is useful for long and repetitive tasks but has to be considered dangerous.

Of course, for a specific application, users are not obliged to instantiate all these qualities. Furthermore, the border between qualities may vary from one application to another one depending on its constraints.

A Gemma-Q has to be defined for each device to control, and has to be implemented on the device, if it has resources to manage the implementation or on a computer connected to it with a safe link (like a serial link). User who will create his own application will also have to define and implement a Gemma-Q, which may or not overlap the device's one.

5. SOFTWARE ARCHITECTURE

A software architecture (Ogor *et al.*, 2001) has also been defined to make remote control of industrial mechanical systems possible. It is based on a set of independent modules running in parallel. On the left side of figure 3, the server side is represented. It is basically composed of three main processes: *Groom* which is in charge of the initial connection, *DeviceManager* which manages the different devices that can be controlled and *ConnectionManager* which manages the different connected clients according to the *ControlAlhorithm* module. This one is choosed by the designer of the system depending on the application: master/slave, priority, timeout... *ToolInterface* are specialized modules that control specific tool.

The right side of the figure 3 represents the client side. Processes are loaded in a web navigator using Java Applet technology. *RemoteClientManager* acts as a router between the *ToolGui* processes and the *NISender* and *NIReceiver* one. A *ToolGui* corresponds to a graphical user interface to control a specific tool: user may send orders and receive information through them. One *ToolGui* is associated to one *ToolInterface* and to one real mechanical system. *NI* processes are used to communicate with the server side.

When a client wants to take control of the system, he will first download applets that will try to establish a communication with the *Groom* using TCP/IP protocol based on socket. If connection is accepted, *Groom* will inform the *ClientManager* and the *DeviceManager*. A set of processes *LocalClientManager* and again *NI* processes will be started to take in charge this new client. If he gets the control of the system (depending on the *ControlAlgorithm*), then he will be able to send commands to the system. In parallel, *Pinger* and *Ponger* processes are launched to observe dynamically the network.

This architecture has been designed as a kernel of services that will take in charge the connection, the management of users and tools, the supervision of the network. Around it, independent

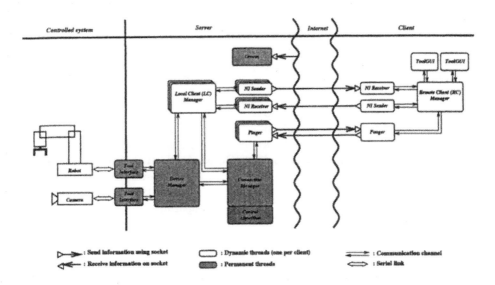

Fig. 3. Software architecture

modules have been defined to manage a specific tool and can be added to the whole system using some "plug and play" mechanism. In the same way, the *ControlAlgorithm* can be chosen in a set of components.

6. APPLICATIONS

Differents applications have been developed using concepts explained in the previous sections. Two of them, the Ericc robot and the penguin motorized camera projects are presented in the following.

6.1 Ericc Robot

This first application (Le Parc *et al.*, n.d.) concerns the control of a robot arm with 5 degrees of freedom. This robot may grasp some small objets with its pliers and move them to other places. A first camera (Sony EVI-D31) is placed in front of the robot. As it is motorized, it can make movements (left, right, up, down) and also has zoom facilities. It enables the distant user to adjust the images to its works. A second camera is placed on the top of the robot to give another point of view of the scene. It is a fixed focus and zoom camera, but it provides an embedded webserver to spread video. An audio module is also availlable.

This robot is used for teaching by professor from our institution and also by colleagues from other universities in France or abroad. It will be available $24^h/24^h$ in some months.

The same development has been made for a rapid prototyping machine and we plan to use it for teaching as well. These two applications show the interest of such a work in the educational context

and its feasibility. Comments from users help us to improve the whole system.

6.2 Penguin motorized camera

Another experiment(Laboratoire LIMI-EA2215 *et al.*, n.d.) has been settled up to control a pan, tilt and zoom camera using the VISCA protocol and has been implemented in a aquarium close to Brest (France) to look at penguins. 4/5 images per second are produced with a size of 192x144 pixels. In this context, the interest of this work is not to protect the camera and its environment in case of network failures (damages are physically not possible), but to check that the kernel of our software is bug-free.

This environment is also used to study network connections: as soon as a client is connected to the server, the ping/pong mechanism is launched and information it collects are stored. After 18 months of uses, statistics show that this motorized webcam has been controlled by 18000 people, coming from all over the World, using modem line, xDSL solutions or Local Area Network. The mean round time trip (time needed for a packet make a Ping/Pong) is respectively around 800ms, 200ms and 150ms. It really shows that controlling such a system is possible with the nowaday Internet technologies and infrastructures.

7. CONCLUSION AND FUTURE DEVELOPMENTS

Our work and the works presented in section 2 show that the remote control of mechanical system over an unpredictable network such as Internet is feasible and will be developed in the close future for tele-teaching, tele-maintenance, tele-expertise or tele-production. A good system may rely on a kernel of services to communicate, to manage people and tools, to study the network and to adapt itself to the quality of the connection. Moreover, the plug and play aspect has to be kept in mind to be able to adapt the whole system to new tools very easily. Such a system may also propose some high level graphical interfaces to help the user to use the whole system and to reduce the gap between the real system and its view through an Internet Navigator (like virtual or augmented reality), and documentation.

In the future, we would like to work in different directions to improve our approach:

- Network sensor: the network is studied using a basic but efficient method: Ping/Pong. It may be enriched by other information like statistical ones or based on the detailed knowledge on the TCP/IP protocol used.
- Multimedia aspects: as the network sensor gives us information about the quality of the connection, we should adapt the quality of the sent images depending on the distant user. A prototype is under development for this purpose.
  More, in the context of robotics, it is generally not useful to transmit all the sounds produced by the system. One of our goal is to apply filters on sounds and to only transmit pertinent information to the user.
- Automatic generation: we have already defined several different *ToolInterface* and *ToolGUI* modules for different tools we are controlling. Instead of redeveloping each time nearly the same code, we would like to propose a model of a tool defined mainly by its interface, its communication protocol, its constraints and then to use compilers to automatically generate the corresponding processes.
- Formal proof: as shown in section 5, our software architecture is based on a set of processes, which are running in parallel and communicating together. The experiment we made seems to show that the whole system is working perfectly. We do not detect any cases of failures (like loss of messages, interblocking...), but to be sure of this result we might use some formal techniques to check the whole application. Some tests have been made using the Spin/Promela model-checker.

All these items are necessary to really insure safety in the context of remote control of mechanical systems over the Internet.

REFERENCES

Benali, A., V. Idasiak and J.G. Fontaine (2001). Remote robot teleoperation via internet. a first approach.. *IEEE International workshop on Robot and Human Interactive Communication* pp. 306–312.

Bicchi, A., A. Coppelli, F. Quarto, L. Rizzo, F. Turchi and A. Balestino (2001). Breaking the lab's walls telelaboratories at the university of pisa. In: *Proceedings of the 2001 IEEE International Conference on Robotics and Automation*. Seoul, Korea.

Burgard, W., A.B. Cremers, D. Fox, D. Hähnel, G. Lakemeyer, D. Schulz, W. Steiner and S. Thrun (1995). The mobile robot rhino.. *AI Magazine*.

Goldberg, K., S. Gentner, C. Sutter and J. Wiegley (2000). The mercury project: a feasibility study for internet robotics.. *IEEE Robotics & Automation Magazine* pp. 35–40.

Harel, D. (1987). *STATECHARTS : A visual formalism for complex systems*. In: *Science of Computer programming*. Vol. 8, 3.

Laboratoire LIMI-EA2215, Oceanopolis and IRVI-Progeneris (n.d.). The penguin project. http://www.oceanopolis.com/visite/visite.htm.

Le Parc, P., P. Ogor, J. Vareille and L. Marcé (n.d.). Robot control from the limi lab. http://similimi.univ-brest.fr.

Oboe, R. and P. Fiorini (1997). Issues on internet-based teleoperation. In: *Syroco 97*. Nantes, France. pp. 611–617.

Ogor, P. (2001). Une architecture générique pour la supervision sûre à distance de machines de production avec Internet. PhD thesis. Université de Bretagne Occidentale.

Ogor, P., P. Le Parc, J. Vareille and L.Marcé (2001). Control a robot on internet. In: *6th IFAC Symposium on Cost Oriented Automation*. Berlin, Germany.

Otmane, S., M. Mallem, A. Kheddar and F. Chavand (2000). Active virtual guide as an apparatus for augmented reality based telemanipulation system on the internet. In: *IEEE Computer Society "33rd Annual Simulation Symposium ANSS 2000"*. Washington D.C., USA. pp. 185 – 191.

Saucy, P. and F.Mondada (2000). Kheponthe web: Open access to a mobile robot on the internet.. *IEEE robotics and automation magazine*.

Simmons, Reid (1998). Xavier : An autonomous mobile robot on the web. In: *In International Workshop On Intelligent Robots and Systems (IROS)*. Victoria, Canada.

Stein, M.R. (1998). Painting on the world wide web : the pumapaint project.. In: *In Proceeding of the IEEE IROS'98 Workshop on Robots on the Web*. pp. Victoria, Canada.

Taylor, Ken and Barney Dalton (1997). Issues in internet telerobotics.. *FSR'97 International Conference on Field and Service Robots*.

Copyright © IFAC Telematics Applications in Automation and Robotics, Espoo, Finland, 2004

A MATLAB/JAVA INTERFACE TO THE MICA WHEELCHAIR

Sven Rönnbäck * David Rosendahl ** Kalevi Hyyppä ***

* CSEE, Luleå University of Technology,sr@sm.luth.se
** Luleå University of Technology, davros-9@student.luth.se
*** CSEE, Luleå University of Technology,kalevi@sm.luth.se

Abstract: In the MICA (Mobile Internet Connected Assistant) project a high tech wheelchair has been controlled remotely with MATLAB/Java software client-server implementations. Sensor servers that read, time stamp and store sensor device measurements to databases, runs concurrently on an embedded PC running Linux.

The network clients used to read sensor values from the servers are written in Java. That gives them portability between different platforms and architectures.

MATLAB client programs runs on stationary computers. These are used to process and visualize collected data from the wheelchair. MATLAB programs are also used to control and make the wheelchair run semi-autonomously.

The combination of MATLAB/Java software is good for fast development, implementation and testing of algorithms in both education and research.

With the client/server approach it is possible to have one computer to run a complex and power consuming control algorithm and another one to handle the GUI. Both computers execute MATLAB programs and they run concurrently and have mutual exchange of data over the Internet.

When a program works properly in MATLAB it can be ported to Java for faster execution speed and portability. *Copyright © 2004 IFAC*

Keywords: Robotic wheelchair, remote-control, telecommands, MATLAB/Java interface

1. INTRODUCTION

Robots and robotic vehicles are usually controlled by software written in C or C++ that runs on local computer systems. MATLAB has nice plotting facilities to visualize data and the fast development loop leads to less bugs in the software. MATLAB is very efficient on matrix calculations and gives the user good insight in variable values.

The MATLAB environment and its current support for the Java [1] programming language has become an excellent environment for fast developing, implementing and testing of algorithms in research and educa-

tion. Java has become a language that is supported in almost all computer environments from tiny-micro controllers to supercomputers and can be run in a span from Mobile phones to Internet Web browsers like the Internet Explorer under Windows.

The sensor clients are written in Java and new the MATLAB algorithms are ported to Java. So the new software is Java based, it would be nice if this software could run in a realtime environment with soft-realtime performance.

Work to use the combination of MATLAB and Java for remote control has been done by Soinio (2003). Where it was demonstrated that it, from MATLAB using Java, is possible to remote control a Lego robot equipped

---

[1] Java technology,http://java.sun.com

with a camera. He used the Java-Lego-Network-Protocol-interface to send control commands to his robot.

Java and MATLAB/SIMULINK can also be used in education to do remote control labs, Sanchez et al. (2002). It demonstrates how this technique can be used for distance teaching and training of operators. Such a system is available 24 hours a day via the computers.

At the Signal and System group at Uppsala University a Java/MATLAB implementation was used on a Linux computer to do some advanced process control on a laboratory-scale plant, Ewerlid et al. (1997).

The MICA software is programmed using a client-server network approach, in a mix of Java, MAT-LAB, C and the C++ programming languages. The wheelchair can with this software be controlled over the Internet using MATLAB. This Paper will mainly describe the MICA server software on the MICA wheelchair.

Similar wheelchair projects as MICA run all over the world, like the one Erwin Prassler describes in Prassler et al. (2001). His wheelchair has an architecture that is similar to the MICA vehicle. It has a field bus for communication, a joystick, an emergency button and a motor controller connected to the common data bus on the wheelchair. It has an on-board computer that runs Linux and lot of hardware connected through serial ports. They do not have a Wave-LAN like in MICA and do no focus on using MATLAB as the developing platform on new algorithms.

The paper is organized as follows. In section 2, the robotic wheelchair is presented as well as the hardware used in it. In section 3, the software is presented, both the client side and the server side. In subsection 3.1 you can find information about how the communication is done between the client/server programs. In section 4, you will find a description on how the Java programs works in MATLAB. In subsection 4.1 the reader can find information on how the synchronization of MATLAB programs over the network is done. In section 5, we present how the wheelchair is operated over the network. Here you will also find a short example of the information flow when some Java classes are used. In section 6 you can read about the results obtained in the tests and the approach used.

2. PRESENTATION OF THE MICA WHEELCHAIR PLATFORM

The work has been done by using a high-tech wheelchair from a local company. We have equipped it with a lot of different sensors and other hardware devices, see Figure 1.

In the figure we can see the wheelchair, the embedded PC and some of the sensors connected to it. Each sensor acts as an independent module that can be

added or removed. We will only need electrical power and interface a new sensor to the embedded PC to be able to use it. On the embedded PC we need to start a server program (software module) that can communicate with the sensor device.

Fig. 1. The MICA wheelchair. The tilted rectangular unit above the seat is a range scanning laser (A). A GPS antenna is mounted on the top (C), near the LazerWay system, LazerWay (Jan). The wireless LAN access-point placed on the seat (B). The rectangular dot seen under the seat to the left of (E) is an inclinometer. Next to the inclinometer a rate gyro is strapdown mounted. At position (D) we have the embedded PC in PC104 format.

The wheelchair is equipped with a CAN bus (Controller Area Network). On the CAN bus small messages are sent and read by other CAN nodes. The communication protocol that runs on the bus was specially made for the wheelchair and its hardware. The wheelchair has a manual control unit with setup facilities and a joystick. With the joystick it is possible for an able user to control the wheelchair with high precision. The vehicle comes with incremental encoders mounted on the shaft of each front wheel. They are used for feedback to the main micro-controller unit.

On the wheelchair several sensor devices are mounted, which are connected to the embedded PC or the CAN bus. Most of the equipment and sensor devices that are connected to the PC104 [2] operates through serial port

---

[2] PC104, http://www.pc104.org,2004

interfaces. To each sensor there exists a program that acts both as a Internet server and as a program that polls the sensor hardware for measurement and stores the information in a local database.

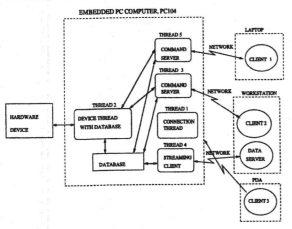

Fig. 2. An example of a device server that runs on the PC104. The clients are arbitrary ones; one can run in MATLAB and another one by a Web browser. The hardware device measurements are logged to the database and controlled using thread 2. Thread 1 is created first and waits for clients to connect. For each connected client a new thread is created, e.q. thread 3 and 5. A streaming thread (thread 4) is created by thread 1 after a request from client 2 and is used to send measurements continuously to the client. Client 3 is just about to connect to the server.A command thread executes commands sent by the connected client and sends the result back to the client. The communication is bidirectional. When a client disconnects the corresponding thread on the PC104 will terminate automatically.

3. SOFTWARE DESCRIPTION

The device servers were programmed in C++ and compiled in the Linux GCC environment. The reason why C++ was chosen on the server side is better hardware access and reuse of code from previous work where the code is based on C++ and the RTAI (Real Time Application Interface). Most of the hardware is connected through serial ports. The current server modules run as different programs and are started separately.

Java servers may be written and compiled with the GNU java (GCJ[3]) compiler to generate pure executable machine code. Java is not a language to use when the software needs hardware access. Java do have support for serial port communication with the Java Communication API [4], so most of the sensors

[3] http://gcc.gnu.org/java/
[4] http://java.sun.com/products/javacomm/

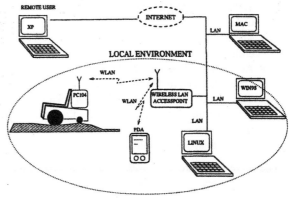

Fig. 3. System architecture overview. The figure illustrates how a computing network nearby the wheelchair can look like. From an end user point of view, the small hand held pocket PC can be used to help a disabled person to handle the wheelchair in difficult situations. The other clients can be research computers with lots of computing power used to do for instance path planning. One client computer can be a remote operator at an emergency Centre or a maintenance company or why not any user on the Internet.

could be accessed with that. If we need pure hardware access like IO pins and AD/DA cards and realtime performance, Java in standard version does not support that, therefore C++ and C is the better choices on the server side.

There are solutions for realtime Java, like the integrated Real-Time specification of Java called JTime, TimeSys (2004). This is the reference implementation for the Real-Time Specification of Java and offers a C/C++/Java cross platform developing environment.

On the client side the language is Java. As Ewerlid et al. (1997) mentioned Java has a lot of advantages, possibilities and features like:

- Java has built in generic network support that makes networked communication easy.
- Java has support for multi-threaded programs, which makes it possible to create advanced programs.
- Java is a rather simple language, and comes with built in garbage collection, so memory leaks is not a big problem. The language is taught to undergraduate students and they can therefore start immediately to create programs in Java.
- Java is a portable language, which makes it possible to run the programs on different types of computers.
- Java is a good language especially on the client side when the programs need a graphical interfaces due to the rich support for graphical user interface (GUI) components.
- Java programs can be compiled to executables programs using an ahead-of-time compiler such

as the GNU (gcj) front end compiler to the GNU C compiler.

- The object orientation, and the built in garbage collect.
- An Java application can be converted into an applet, and run in a Web-browser.
- The possibility to extend the language with other programming languages. By using the Java native interface (JNI) it is possible to develop libraries to access peripheral devices.

The server software talks with the Java clients using TCP/IP sockets over wireless LAN.

3.1 Client-Server Communication

The data transfer over the network is done using pure readable ASCII text (American Standard Code for Information Interchange), by this choice the transferred data can easily be analyzed and debugged by a human. If multiple lines are sent the transfer starts with a BEGIN command and ends with an END command. The received strings are parsed and put into a Java object. The data can also be sent in a binary format, this is more efficient with respect to time and bandwidth but it will make it harder to debug the information. The binary transfer was specially introduced as an option to transfer laser scans faster to the laser clients due to the huge amount of data that needs to be sent. It also reduced the CPU usage a lot when the software does not need to convert integer numbers to ASCII text.

3.2 Description of the Software on the Server Side

The servers are written in C and C++ and all of them have the same structure. The idea has been to make the code portable between real time Linux and normal Linux. The structure is such that the actual scheduling and frequency of a thread is handled by a central node. This makes it easier to change the operating system, to for instance QNX, since only the scheduling node has to be ported. Mutual exclusion of data is done using a semaphore class. Only this class has to be updated with a new semaphore structure, to work on a new realtime OS.

The sensor thread collects measurements from a sensor and puts them in a database. All measurements get an increasing index and are time stamped with the current system time of the PC104. When the measurements from all the sensors are time stamped in the same way, it is possible to actually do some interesting sensor fusion of the information provided.

The exchange of the information between the threads is secured using a semaphore class written based on POSIX 1003.1b [5] semaphores. Our semaphore

---

[5] http://standards.ieee.org/regauth/posix/

class contains some abilities to debug the use of the semaphore object.

Another thread on the sensor side waits for clients to connect to a specific given socket port. As a new client connects yet another thread is created to assist the client requests. By this way it is possible for a lot of clients to connect to the same server and the server can serve them all.

For example, in the CAN server, a connected client can put itself as a master and then commands from other clients will be ignored. This option is implemented for safety reasons so that the automatic control can be overtaken by a human with a remote control unit.

3.3 The Java Clients

All the clients are written in Java. When a program is written in Java it is very easy to use the code in almost any environment or modern computer architecture. The data, returned to a client, is encapsulated in an object. That holds an index with incremental values used to identify the measurement, the time when the measurement was made and the actual measurement.

In previous works the clients were written with the Mex-file support in MATLAB. When this was done the clients were locked to the MATLAB environment and hence the code is more bounded to the computer architecture.

4. JAVA IN THE MATLAB ENVIRONMENT

One nice property when Java objects are called in MATLAB is the possibility to return matrices of both integer and double type. For example a double array return type will result in a matrix in MATLAB. In the same way if a Java function is called with a MATLAB matrix array as an argument it will come as a two dimensional double array in Java. The Java classes are controlled using M-files in MATLAB. Functions to initiate, call the functions inside a class is efficiently implemented using M-files.

4.1 MATLAB Client Synchronization Over The Network

A network pipe was implemented with the functions Put and Get. This makes it possible to send a matrix, vector or scalar from one MATLAB client to another. One client starts the server and waits for a client to connect, the pipe-client and the pipe-server act as a pair and it is only possible to send information one way. If we want to send information both ways we need a pipe-server and a pipe-client on each side, see Figure 4. In the figure there are two MATLAB programs running, and they exchange information with two data pipes.

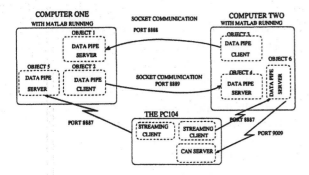

Fig. 4. An example where two MATLAB programs are running. They use the network data pipe implemented in Java to exchange matrix information. The data pipe client(object one) sends matrix data to the server (object three). In the current example a answer can be sent back by using the object two and four. The PC104 streams laser scan data with a set frequency to object five so that MATLAB computer one who processes that information and sends the result to computer two. From the PC104, odometric data and gyro data are streamed to object six in MATLAB computer two which runs a control algorithm and sends control commands to the PC104.

5. SOFT REALTIME OPERATIONS OF THE WHEELCHAIR FROM MATLAB

When we talk about robots and robotic system we want to control them in some way, we can talk about telecontrol and telecommands. A very simple example in this case of telecontrol is just to put the steering joystick on the wheelchair on another system, like in a MATLAB program or on Pocket PC computer. But this comes with risks as well, the longer distance there is to the robot the more risky it will become due to the transportation time of the signals. One way to bypass this problem is to implement telecommands. As the developing process of new algorithms is done in MATLAB the dynamics of the mobile robot is restricted to slow moments and low speed. This is not a problem as long as we run the system slow mode. A discrete time period of 0.5 seconds has been used on some MATLAB implementations. Most of that time are consumed in the polling process of new measurements over the network. The MICA wheelchair has support for telecommands. Telecommands have been implemented by for instance Högström et al. (1995); Forsberg et al. (1998); Wernersson et al. (1995); Schilling and Roth (1999); Schilling and Vernet (2002). Here is list of commands found in the MICA system:

- Set the speed of the wheelchair
- Turn the wheelchair
- Set turn rate with use of the rate gyro
- Set the heading
- Calibrate inertial navigation system
- Take a laser scan
- Drive the vehicle X meters with velocity ($f_v(s)$) and heading ($f_\theta(s)$), $s \in [0..X]$

- Follow a path segment of length X with coordinates $[x_w(s), y_w(s)]^T = f_p(s)$, $s \in [0..X]$
- Stop the wheelchair or Emergency-break
- Add, change, remove, and execute way-points
- Run down the corridor
- Follow a wall

A CAN client has been implemented in Java, that has some driving commands that can be used to remote control the wheelchair. It uses a special CAN protocol called Bore-Can[6] to send CAN messages to the wheelchair.

5.1 An Example of Remote Wheelchair Operations

An example of a MATLAB script for remote operation and data collection. The script is written with a special notation to indicate the return values from functions or the types of a variables. A similar MATLAB script as the one below was used to create Figure 5.

```
1  >> oControl=Control('rullstol.sm.luth.se')
2  >> oControl.rSetVelocityAndOmega(0,2.5*pi/180)
6  >> rTime=oControl.rGetSystemTime
4  >> input('Press a key')
5  >> oControl.bStop(true)
6  >> rDT=roControl.rGetSystemTime-rTime;
7  >> aoPos=oControl.aoGetPositionByTime(rTime-rDT,rDT)
8  >> oLaser=Laser('rullstol.sm.luth.se')
9  >> aoScan=oLaser.aoGetScanByTime(rTime-rDT,rDT)
10 >> oGyro=Gyro('rullstol.sm.luth.se')
11 >> aoRate=oGyro.aoGetRateByTime(rTime-rDT,rDT)
12 >>   % plotting and analyzing of data.
           % Find the door and calculate route.
13 >> anPos=oControl.anMoveToPos(aarRoute)
14 >> oControl.anRun(anPos)
15 >> oControl.disconnect
```

On line 1-2 the controller client are used to set the heading and the velocity of the wheelchair. Line 3 sets a angular velocity of $2.5°/s$. Line 4 waits for an input from the user. Line 5 stops the movement of the wheelchair. The system time is returned to a variable in line 4. The the time length of the run is calculated on line 6. Line 6 polls the server for position objects during the scanning. On line 7 we create a range laser client object from the Laser class. The laser server are polled in line 8. On line 9-10 a gyro client is created. On line 12 we do the calculation and analyzing on the received data. Line 13 adds an route to to the controller. On line 14 tells the controller to run the wheelchair through the way-points $anPos$.

6. RESULTS

It has been shown in this work that the combination of Java and MATLAB is a good environment for fast developing and testing of new algorithms in the robotics field. The MATLAB environment gives the user full insight in variable values and extensive plotting facilities, and therefore it provides fast debugging of the code for the user.

---

[6] Bore is taken from Boden Rehab, the company that designed the wheelchair. Bore-Can is the special protocol for CAN messages on the embedded CAN bus in the wheelchair.

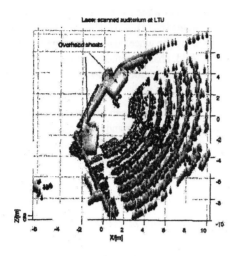

Fig. 5. A contour plot of an auditorium at Luleå University of Technology. The wheelchair was programmed to scan the environment with a SICK laser. The dark areas is near the floor and the light areas are closer to the roof. Two overhead sheets are marked in the figure.

The Java programming language makes it possible to reuse the clients as standalone programs and the feasibility to make applets that can run under Web-browsers and other applet viewers. The communication between the processes, both clients and servers uses the network, this makes it possible to distribute information and computer power.

Under the developing process of new algorithms we write distributed programs that has several clients that use and process data over an Internet connected network. This means that a lot of that is shuffled and send between nodes over a network that has limited bandwidth. If data is send between threads using the local-host, the bandwidth of the data transfer is set by the operating system. The localhost network has no bottle neck set by Ethernet or Wave-LAN (IEEE802.11b). If programs and threads run on the localhost the bandwidth is very high in comparison with Wave-LAN and the data transfer is not affected by time jitter and the network performance. On Wave-LAN the communication is very dependent on the environment because it is based on radio waves and can therefore it can easily be jammed and have net disturbances that affect the data transfer.

7. CONCLUSIONS

It is possible to use Java as a programming language in the MATLAB environment. A robotic wheelchair has been remote controlled with simple control algorithms written in MATLAB. The MATLAB environment is suitable for extensive algorithms like map building, localization, path finding, wall extraction and obstacle detection. As the MATLAB algorithms work properly, they are ported to the Java language for more efficient

computing (especially nested loops). This gives the possibility to keep them in MATLAB or run them on any network computer client or put them as a separately program on the PC104. For example: Hough's transform is used to extract walls and lines from laser scans. This algorithm is implemented in MATLAB and polls the SICK laser server for laser scans. The Hough transform code should be ported to Java and then used to write a Hough transform server that runs on the PC104.

8. ACKNOWLEDGMENTS

The MICA project is partly funded by Interreg IIIA Nord. The writers want to thank the robotics project students, year 2003.

REFERENCES

Ove Ewerlid, Claes Tidestad, and Mikael Sternad. Realtime control using matlab and java. *Nordic MATLAB Conference*, 1997.

Johan Forsberg, Ulf Larsson, and Åke Wernersson. Tele-commands for mobile robot navigation using range measurements. *Paper in PhD thesis:Mobile Robot Navigation Using Non-Contact Sensors, Johan Forsberg, ISSN:1402-1544, Luleå University of Technology*, 1998.

Tomas Högström, Jonas Nygårds, Johan Forsberg, and Åke Wernersson. Telecommands for remotely operated vehicles. *IFAC, Intelligent Autonomous Vehicles*, 1995.

LazerWay. Lazerway. http://www.lazerway.com, 2004 Jan.

Erwin Prassler, Jens Scholz, and Paolo Fiorini. An autonomous vehicle for people with motor disabilities. *IEEE Rob. Automat. Mag.*, 7:38–45, 2001.

J. Sanchez, F. Morilla, S. Dormido, J. Aranda, and P. Ruiperez. Virtual and remote control labs using java: a qualitative approach. *IEEE Control Systems Magazine*, 22(2):8–20, Apr 2002.

K.J. Schilling and H. Roth. Control interfaces for tele-operated mobile robots. In *Proceedings on Emerging Technologies and Factory Automation ETFA*, volume 2, pages 1399–1403. IEEE, Oct 1999.

K.J Schilling and M.P. Vernet. Remotely controlled experiments with mobile robots. In *Proceedings of the Thirty-Fourth Southeastern Symposium on System Theory*, pages 71–74. IEEE, March 2002.

Asmo Soinio. A lego-robot with camera controlled by matlab. http://www.abo.fi/fak/ktf/rt/robot/, 2003.

TimeSys. Jtime, java technology for embedded and real-time developement. http://www.timesys.com, Jan 2004.

Å. Wernersson, M. Blomquist, J. Nygårds, and T. Högström. Telecommands for semiautonomous operations. In *Proc. Telemanipulator and Telepresence Technologies*, volume 2351, pages 2–12. SPIE, 1995.

Copyright © IFAC Telematics Applications in Automation
and Robotics, Espoo, Finland, 2004

ELSEVIER

IFAC

PUBLICATIONS

www.elsevier.com/locate/ifac

CELLULAR NETWORK TELECONTROLLED ROBOT VEHICLE

Alar Kuusik, Takeo Hisada, Satoshi Suzuki, Katsuhisa Furuta

Tokyo Denki University
College of Science and Engineering,
Department of Computers and Systems Engineering.
Hatoyama, Saitama 350-0394, JAPAN
{Kuusik, Hisada}@furutalab.k.dendai.ac.jp

Abstract: Current paper presents development results of 3G cellular network based telecontrol system for a wheeled mobile robot. An embedded picture driven teleoperation system was designed to control the robot vehicle through the web-enabled cellular phones. Several issues of creating fast enough and flexible hardware and software solution are addressed. Experimental results describing control speed and command latency through a web-enabled 3G cellular phone are presented. *Copyright © 2004 IFAC*

Keywords: mobile robots, telecontrol, embedded systems, robot control, communication channels, computer controlled systems, control system design

1.INTRODUCTION

Mobile robot vehicles dedicated for remote manipulation or environment observation are under rapid development for wide range of research- and commercial purposes. There are several tele-manipulation devices in real use for explosives disposal (Hicom Security). One important issue of robot teleoperation is choosing most appropriate communication channel between operator and device. For unique and precise tasks skillful operators have to be used and long range teleoperation established.

Recently many prototypes of wireless control robots using IEEE802.11x (WiFi) have been designed (Cremean, et. al. 2002; Potgieter et al., 2002). However, sometimes it is complicated to set up WiFi network, more extended control range is required or sophisticated data analysis have to be performed far away. Cellular network based telecontrol is a good choice in such cases, especially if 3G high speed networks can be used. In Japan, where 3G networks are widely available, such control experiments can be easily performed. Only few attempts have been made so far to control robots through cellular network using WAP (Gerogiannakis and Sgouros, 2002). Naturally, 2G communication bandwidth is insufficient for

picture transmission essential for telemanipulation. The goal of the present project was to develop an experimental mobile vehicle control system with picture transmission capabilities for teleoperation through 3G cellular networks and test performance of the solution.

Most of the existing experimental mobile robots designed at research institutions over the world use regular PC motherboards, or (rack mount) PC-s for control and communication (Yasutake *et al.*, 2001, Han *et. al.* 2002, Tomatis *et al.*, 2002). That solution results heavy weight, large dimensions of device and power consumption of at least 100 watts. Such control hardware is quite unsuitable for practically useful robotic devices. Additionally, robots controlled with conventional PC type hardware and general use desktop software have insufficient reliability (Tomatis *et al.*, 2002). One goal of the present project was to develop reliable embedded control system for robotic devices that employs low power consumption (preferably below 5W), solid construction, small size and uses some existing communication protocol. Because of resource constraints the solution had to be realized from off-the-self components for controller and remote terminal using non proprietary software solutions.

2. WHEELED MOBILE ROBOT

The experimental wheeled mobile robot (WMR) designed at Tokyo Denki University is shown in Fig. 1. Controller with PCCARD type 3G cellular modem can be seen on the top left side of WMR. Weight of WMR is 47kg (long run batteries), its moving scheme is "differential type" – two motorized wheels can be driven in both direction by pulse-width-modulated current control. The front wheel is a passive one. Custom designed pulse width modulators offer 10bit resolution for torque control. Internal wheel pulse counters for position measurements have resolution of 0.02 degrees.

Fig. 1. Wheeled mobile robot.

2.1 Movement control of WMR

The movement control of vehicle can be understood following Fig. 2.

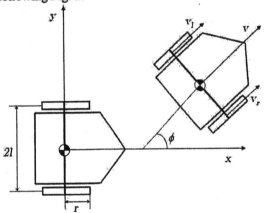

Fig. 2. WMR turning scheme.

The model of the vehicle is described as follows:

$$v_r = r\,\dot{\theta}_r, v_l = r\,\dot{\theta}_l \tag{1}$$

$$\frac{v_r - v_l}{2l} = \dot{\phi} \tag{2}$$

$$\frac{v_r + v_l}{2} = v \tag{3}$$

$$\tau_* = I_w \ddot{\theta}_* + c_w \dot{\theta}_* \tag{4}$$

where v_l and v_r - velocities of left and right wheel,
r – radius of wheels (here: 6.5 cm),
l – length from center of gravity to a wheel (here: 24cm),
ϕ – turning angle of the vehicle with initial value of ϕ_0,
v – velocity of the center of gravity of the vehicle,
θ_r and θ_l - rotation angles of wheels,
τ_*- driving torque of motor for corresponding wheel,
I_w – moment of inertia of a wheel, (here: 0.02 kg*m²),
c_w – viscous friction factor (here: 0.17 N*m/s).

The commands from operator are desired angle ϕ and velocity v. From (2) and (3) and known ϕ_0 the given angle ϕ and speed v determine the reference velocities v_l and v_r. There is a feedback control system implemented to track the velocities of wheels v_l, v_r to reference velocities v_l, v_r. Block scheme of the control loop is shown on Figure 3. The function block F transforms the reference angle ϕ and velocity v to the actual velocities of wheels. i_l and i_r are currents of left and right motor that are proportional to driving torques τ_l and τ_r.

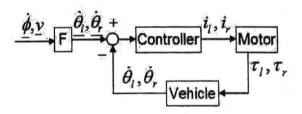

Fig. 3. Block scheme of feedback control.

The PID controller is used. From simulation results the chosen control system realizes tracking to the operator's commands with accuracy of 2%. The implemented movement control system requires onboard computer with computing power to achieve torque calculation in sufficiently short time interval.

2.2 Requirements for control stability

Required real time performance (needful computational power) determined by essential task switching frequency of controller depends on the maximum stable execution period of the control loop.

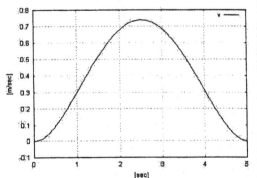

Fig. 4. Stable WMR control loop, sampling time 53ms, tracking error less than 2%.

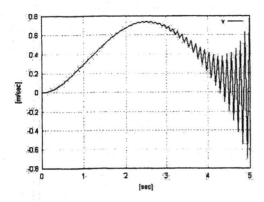

Fig. 5. Unstable control loop, sampling time 54 ms.

Fig. 4 and 5 demonstrate simulation results of large sampling period for particular mechanical system showing the emergence of instability. Desired speed curve is a sine wave with amplitude of 0.75 m/s. Fig. 4. has the loop execution period of 53ms (stable) and 54ms (unstable) Fig. Therefore the CPU of WMR has to be able to execute control law with a period of less than 50ms to maintain stability.

3.REMOTE USER INTERFACE

Finding a suitable, preferably off-the-self remote control device is a key issue for rapid realization of all kind of custom remote control systems. Requirements for particular handheld remote device were: cellular modem support, large enough display (min 2"), joystick or touchscreen HID, Internet compatibility. The second main issue is finding a reasonable and flexible software solution for user interface development.

3.1 Remote user terminal hardware.

Considered remote control device options were (notebook) PC, cell phone, PDA supporting PCCARD or CF type cellular modems. PDA has been used for robot control several times before (Lu et al., 2001; Girson, 2002). However, recently PDA like cellular phones with large display and full size touchscreen became available. Using such devices provided by cell company itself is safer from network compatibility point of view. Moreover, subsidization those devices is also important. Chosen device of NTT DoCoMo FOMA SH2101V has 2.5" display and Java enabled browser built in. Main disadvantage of the device appeared to be short battery lifetime. NTT was chosen as available 3G network at campus area.

3.2 Web-based remote control.

Web-based control is probably the only reasonable solution for interfacing experimental unique devices with large volume consumer devices like cellular phones. Fortunately, this web-based control is widely

tested since Goldberg introduced his robotic arm ten years ago (Goldberg, et al., 1995). Nowadays the method is used in industrial control proving sufficient dependability. Many teleoperated robotic systems employ web-based control solutions (Hu et al., 2001; Chen and Luo, 1997; Brady and Tarn, 1998; Taylor and Dalton, 1997; Kim et al., 2002; Safaric et al., 2001). However, there are quality-of-service problems related to the web-based telecontrol. So called "cognitive fatigue" is a consequence of controlling every single movement (Murphy, 2000). Improvements can be achieved by making controller intelligent and using higher level commands e.g. "return to initial position", "go to X,Y" (Marín et al., 2002). By present WMR it requires some additional embedded computing power but it is easy to realize using chosen web-based control technique. It is easy to show and update a map of operation area through browser and receive destination coordinates as a macro command. The macro command approach is also less sensitive to variable Internet delays (Kress et al., 2001).

Internet latency is second main disadvantage of web-based control. Actual delay can estimated be by round trip delay of ping packages (Oboe and Fiorini, 1998). Experiments in NTT WCDMA network in different modes (circuit, packet switching), presented later, showed that ping delay can be up to one second . The shortest transmission delay achieved in particular network (through network provider Internet access address) was close to maximal allowed delay of 200ms of effective teleoperation (Thompson et al., 1993). Some formal methods are recommended to improve delay handling (Mirfakhrai and Payandeh, 2002) and an idea of *virtual machine* can be used (Han et al., 2001; Marín et al., 2002). Unfortunately those approaches require additional operator's console computational power unavailable for cell phones. Conclusively, the implemented movement control software runs entirely on onboard computer and *move and wait* (Sherindan, 1992) strategy was adapted for remote operation.

4.CONTROLLER HARDWARE DESIGN

Requirements to embedded controller hardware were: off-the-self device with power consumption less than 5W; enough computing capabilities to handle simultaneous picture frame grabbing and control law execution; embedded interfaces for cellular modems and WiFi cards (PCCARD or CF card), external ports for motor control and CCD camera, watchdog presence.

Simultaneous execution of described "soft" realtime tasks require at least 32 bit processor. Similar but simpler robot control task and web page serving task have been performed separately on two Pentium II 266 MHz type PCs (Goldberg et al., 2000). The present goal was executing both tasks on a single controller with similar computational power.

4.1 Selecting an embedded CPU.

Powerful embedded computers with open IO interfaces (some kind of parallel bus) have PIII, Mobile Athlon/Celeron or VIA C3 CPUs with computing power about 1000-2000 MIPS. Such devices consume at least 20-30W of energy exceeding design target. As target solution following low power 32 bit around 1W power consumption CPU types were investigated: Intel PXA255 (400MHz, Xscale, ARM), NEC VR4181A (131MHz, MIPS), AMD Alchemy Au1500 (400MHz, MIPS), NSC/AMD Geode Sc1100 (266MHz, x86), Hitachi SH7709S (167MHz, SH3), Motorola MCF5249 (140MHz, 68K/ColdFire). Performance comparison of those CPUs is presented in the Table 1 (EEMBC).

Table 1 Comparison of 1W class embedded CPUs

CPU	Energy consumption, mW	Thrystone 2.1 MIPS	FPU/ INT Div/ Mult	Disadvantages
PXA255	450	480	N	No onchip USB host
VR4181A	500	155	N	
Au1500	700	480	N/D	
SC1100	1000	245	Y	No onchip PCMCIA
SH7709S	650	220	N/M	
MCF5249	185	125	N/D	No USB

Most of embedded devices lack of floating point unit as shown in column 4. FPU was considered quite essential for chosen control law. XScale is getting popular in embedded world and many off the self boards can be found (Advantech, Kontron). Unfortunately XScale does not have host USB controller essential for connecting USB CCD cameras. Most widely offered embedded solutions still use x86 processor architecture.

Embedded computers based on Geode series from AMD appeared to be best solution for WMR control present moment. Also, lately announced Alchemy series from AMD seems promising for designing flexible embedded controllers for robots, especially because the true PCI interface is present (Au1000). Unfortunately only few manufacturers so far offer Alchemy based embedded computers (Ampro, Laser5).

SC1100 Geode x86 was chosen as most suitable CPU for particular application. It has midrange computing power and fully featured FPU. The x86 architecture still has best software knowledge base. A suitable and inexpensive embedded computer with dual CardBus controller and was found (Routerboard). Dependability measures are not provided for the board but experiments showed good results. In the further MTBF could be estimated by using RelCalc or any other reliability calculation program. Also, software may be ported to a modern MIPS architecture CPU.

Custom motor control circuit board was designed using FPGA (reprogrammable flash logic IC) that collects movement information from wheels (counter input) and generates PWM signals for both motors. MTBF of the custom board calculated by RelCalc Version 5.0-217F2 (MIL-HDBK-217F standard) is more than 500000 hours. Current solution has single USB CCD camera, audio signal from the camera can be used in the further considered useful for real teleoperation. Cellular modem or WiFi network card can be attached to dual PCCARD slot.

5. WMR SOFTWARE

5.1 Multiagent control.

The control system is designed to support modes: Single Operator Single Robot (SOSR), Multiple Operator Single Robot (MOSR) and Single Operator Multiple Robot (SOMR) according to classification proposed by Tanie (Chong et al., 2000). MOSR allows several operators to be connected to the robot (to the onboard web server) on the same time. Each operator can focus on a single task of controlling the vehicle or environment monitoring. SOMR configuration allows controlling several robots by a single operator. In such case the operator can control virtual gravity center of a set of robots. This solution useful for moving heavy items, for example. In SOMR mode one robot is set up as a master receiving commands from operator, other robots can establish connection to master device to get the information for given task.

5.2 WMR software implementation.

WMR control page on screen of SH2101V 3G cellular phone is shown in Fig. 6.

Fig. 6. User GUI on screen of web cellphone.

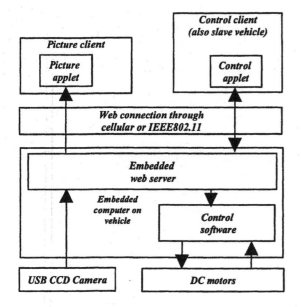

Fig. 7. Control scheme of robot vehicle.

Fig. 6 shows direct mode control (inputs are desired speed and direction). A circle on right side of screen represents "driving wheel" and online camera picture is shown on the left side. The macro (position) mode control can be used replacing control wheel with a map of room. Picture frame shows USB camera picture with resolution of 320x240 dots. That resolution is used to shrink file size. FTP upload method is used in web-server. Experiments showed that updating a picture frame of 3kB takes 2 seconds on screen of SH2101V (64kbps upload speed of WMR modem). For the same reason frame rate of USB CCD camera is decreased down to 5fps to preserve CPU resources. In the further Quicktime movie format should be evaluated for smooth picture transmission. Quicktime is used for Videophone call application available in Japan. .

Structure of WMR application software running on vehicle computer is shown on Fig. 7. Embedded web server (Boa) provides user with main WMR page. Two different pages are available with and without Java support depending communication channel throughput (WiFi or cellular). PPP server and client software manages incoming and outgoing modem communication.

5.3 Operating system of WMR

Following embedded OS were considered: WinCE, FreeBSD and Linux. Linux as most well supported solution was chosen. Also, 2.6 series kernel has sufficient soft real time properties – task scheduler works now with time slices are below 10ms. Linux has among others the widest range of supported hardware including Geode CPU and wireless network cards. As one of the most widely used embedded Linux distribution - Busybox was chosen (Busybox). Performed tests showed that 233MHz Geode MMX

CPU was able to execute motor control loop with frequency of 100Hz and serve 5fps picture to several clients. Size of complete Linux distribution is 20MB.

6. EXPERIMENTAL RESULTS

Experimental results describing communication speed and latency can be presented. In tests maximum download and upload speed of both cellular modems was 384kbps and 64kbps respectively (packet switching), both modems were in the same base station radio coverage area. Direct dial in PPP is supported only through 64kbps circuit switching channel. Measured latencies of 56bytes ping packet between WMR and remote device are presented in Table 2:

Table 2 PING packet latency

Communication interface	Latency, ms	Remote device
100 Mb Ethernet (local)	0.5	PC
64kbps PPP (circuit sw)	900	SH2101V
64kbps (circuit sw, connected through ISP)	320	SH2101V
Packet sw, 64kbps upload, 384kbps download	180	SH2101V

As seen, direct 64kbps PPP dial in connection between two modems (circuit switching) results much longer latency comparing with connection established through Internet access point of the cellular company (both circuit and packet switching).

WMR was controlled through 100Mb wired Ethernet and cellular network. Command delay over Ethernet, using Opera browser on 1.5GHz PC, latency of command reaching motor drivers – delay less than 0.2s. Delay was mainly caused by control program. Picture refresh in browser: instant (5fps). Cellular communication using SH2101V cellular phone: motor command latency 1.5s, picture update ca 3s (frames skipped).

7. CONCLUSIONS AND FURTHER WORK

A flexible, cost and power effective wireless control solution for an experimental wheeled mobile robot was developed. Full system including robot itself was built for less than 12 months. The picture driven remote operation of the robot can be performed through IEEE802.11 and 3G cellular networks. PCs, PDAs or cellular phones with web-browser can be used as console devices. Flexible software supporting different telecontrol modes is entirely based on freeware. Experiments in a WCDMA cellular network showed that quality of service presently offered to consumers is insufficient for professional teleoperation. In the further latency issues will be more deeply investigated and perhaps some improvements can be made. It is planned to test faster modems and streaming video solutions used by Japanese Videophone applications.

ACKNOWLEDGEMENTS

This paper presents research supported by Japan Society for the Promotion Science (JSPS) 2003 postdoctoral fellowship program.

REFERENCES

Ampro. http://www.ampro.com

Brady, K. and T. J. Tarn (1998). Internet-Based Remote Teleoperation. In *Proc. IEEE Int. Conf. Robot. Automation*, pp. 65-70.

BusyBox. http://www.busybox.net

Chen, T.M. and R. C. Luo (1997). Remote Supervisory Control of An Autonomous Mobile Robot Via World Wide Web. In *Proc. IEEE Int. Symposium on Industrial Electronics*, **vol. 1**, pp. ss60-ss64.

Chong, N., T. Kotoku, K. Ohba, K. Komoriya, N. Matsuhira, and K.Tanie (2000). Remote coordinated controls in multiple telerobot cooperation. In *Proc. IEEE Int. Conf. Robotics and Automation*, **vol. 4**, pp. 3138–3143.

Cremean, L., W.B. Dunbar, D. van Gogh, J. Hickey, E. Klavins, J. Meltzer and R. M. Murray (2001). The Caltech Multi-Vehicle Wireless Testbed. *Conference on Decision and Control (CDC)*.

EEMBC. http://www.eembc.org

Gerogiannakis, S., N. Sgouros (2002). Mobile Robot Teleoperation using Mobile Phones. University of Piraeus

Girson, A. (2002). High-Mobility Tactical Micro-Robot Enters the Field with InHand's Fingertip Technology. *Design Strategies and Methodologies*, **vol. 1**, no. 1.

Goldberg, K., M. Maschna, S. Gentner (1995). Desktop Teleoperation Via The WWW. In *Proceedings of the IEEE International Conference on Robotics and Automation*, pp. 654-659, Japan.

Goldberg, K., B. Chen, R. Solomon, S. Bui, B. Farzin, J. Heitler, D. Poon, and G. Smith (2000). Collaborative teleoperation via the Internet. In *Proc. IEEE Int. Conf. Robotics and Automation*, **vol. 2**, 2000, pp. 2019–2024.

Han, K.-H., S. Kim, Y.-J. Kim, and J.-H. Kim (2001). Internet Control Architecture for Internet-Based Personal Robot. *Autonomous Robots Journal*, **vol. 10**, no. 2, pp. 135-147. Kluwer Academic Publishers.

Han, K.-H., Y. Kim, J. Kim, and S. Hsia (2002). Internet control of personal robot between KAIST and UC Davis. In *Proc. IEEE Int. Conf. Robotics and Automation*, **vol. 2**, pp. 2184–2189.

Hicom Security. http://www.hicomsecurity.com, MR-5, An Explosives Disposal Robot

Hu, H., L. Yu, P. Tsui, and Q. Zhou (2001). Internet-based robotic systems for teleoperation. In *Assembly Automat. J.*, **vol. 21**, no. 2, pp. 143-151.

Kim, J., B. Choi, S. Park, K. Kim, and S. Ko (2002). Remote control system using real-time MPEG-4 streaming technology for mobile robot. In *Dig. Tech. Papers IEEE Int. Conf. Consumer Electronics*, pp. 200–201.

Kontron. http://www.kontron.com

Kress, R.L., W.R. Hamel, P. Murray, and K. Bills (2001). Control Strategies for Teleoperated Internet Assembly, In IEEE/ASME Transactions on Mechatronics, **Vol. 6**, No. 6.

Laser5. Http://www.laser5.co.jp

Lu, W., J. Castellanes and O. Rodrigues (2001). http://www.ncart.scs.ryerson.ca/~wlu/thesis/pdam ax.pdf

Marín, R., P. J. Sanz, A. P. del Pobil (2002). A Predictive Interface Based on Virtual and Augmented Reality for Task Specification in a Web Telerobotic System, Proceedings of the 2002 IEEE/RSJ Intl. Conference on Intelligent Robots and Systems EPFL, pp. 3005-3010, Lausanne

Mirfakhrai, T. and S. Payandeh (2002). A delay prediction approach for teleoperation over the internet. In *IEEE International Conference on Robotics and Automation (ICRA)*.

Murphy, R.R. (2000), Introduction to AI Robotics. The MIT Press.

Oboe, R., and P. Fiorini (1998). A design and control environment for internetbased telerobotics. In *Int. J. Robot. Res.*, pp. 443–449.

Potgieter, J., G. Bright, O. Diegel, S. Tlale (2002). Internet Control Of A Domestic Robot Using A Wireless Lan. In *Proc. Of Australian Conference on Robotics and Automation*, pp. 212-215, Auckland.

Routerboard. http://www.routerboard.com

Safaric, R., M. Debevc, R. Parkin, and S. Uran (2001). Telerobotics experiments via Internet. In *IEEE Trans. Ind. Electron.*, **vol. 48**, pp. 424–431.

Sherindan, T. (1992), Telerobotics, Automation, and Human Supervisory Control. MIT Press.

Taylor, K. and Dalton, B. (1997). Issues in Internet telerobotics. In *Int. Conf. on Field and Service Robotics*.

Thompson, D., B. Burks, and S. Killough (1993). Remote excavation using the telerobotic small emplacement excavator. In *Proc. 5th ANS Int. Topical Meeting Robotics and Remote Systems*, pp. 465–470. Knox, TN.

Tomatis, N., G. Terrien, R. Piguet, D. Burnier, S. Bouabdallah, R. Siegwart (2002). Design and System Integration for the Expo.02 Robot asl.epfl.ch/aslInternalWeb/ASL/publications/uploadedFiles/iros02.pdf

Yasutake T., Shoichi I., Shujiro I., Kouichi H., Yutaka, K. and Minoru A. (2001), http://www.er.ams.eng.osaka-u.ac.jp/robocup/trackies/team01/trackies01.pdf

Copyright © IFAC Telematics Applications in Automation and Robotics, Espoo, Finland, 2004

ELSEVIER

IFAC
PUBLICATIONS
www.elsevier.com/locate/ifac

WIRELESS TELEOPERATION OF AN ASSISTIVE ROBOT BY PDA

R. Correal A. Jardon A. Gimenez C. Balaguer

Robotics Lab, University Carlos III of Madrid, Spain

Abstract: The assistance of disabled, elderly and persons with special needs become to be one of the most important service application of the robotic systems in the near future. Humans care and service demands an innovative robotic solutions to make easier the day-life of these people in home, workplace and institutional care environments. The MATS project is developing a new concept of climbing robot for this type of service applications. The service robot MATS helps disable and elderly people in their day life activities in common living environment like kitchen, bathroom, bedroom, etc. In this way the quality of life of the important part of population improves toward their social integration. This new prototype has new abilities like climb from one wall to another or from the table to the wheelchair, and at the same time to be attached and move with the wheelchair. The robot is totally autonomous and needs only power supply to be operated. This paper presents the distributed software architecture and the concept design of the HMI which handles the robotic system. *Copyright © 2004 IFAC*

Keywords: rehabilitation robotics, wireless, teleoperation

1. INTRODUCTION

During the last years the rehabilitation technology is developing towards more flexible and adaptable robotic systems. These robots aim at supporting disable and elderly people with special needs in their home environment. Furthermore, most advanced countries are becoming to be aging societies, and the percentage of people with special needs is already significant and due to grow. There have been very interesting developments in this field, such as Paro (K. Wada, 2002), a robot which provide psychological and social effects to human beings, or Dexter (L. Zollo, 2001) and RAID (G. Bolmsjo, 1995), which are mounted on a wheelchair and helps in welfare tasks to the disable people. Another interesting robots is Handy 1 (Topping, 2002) and MANUS (Kwee, 1997). These systems are a commercially available robotic system capable of assisting the most severely disabled people in self-feeding, and personal hygiene tasks.

Along this road, the MATS consortium, leaded by Straffordshire University (UK) and composed of other nine partners, is developing a robotic system that can climb the different planes of the environment (wall, table, bath,...) in order to perform numerous personal care and service applications such as kitchen plates and cup manipulation, toothbrush and toothpaste coordinated manipulation in the bathroom and entertainment games in the living room (G. Bolmsjo, 2002). The revolutionary design concept of the MATS system will provide flexible mechatronic assistive technology with functions and features satisfying the expressed desires and priorities of a broad spectrum of potential users.

This paper presents the new concept in the rehabilitation robotics (K. Kawamura, 1995). The main advantage of the MATS robot concept is the combination of both, static and moving systems, into one climbing robot. The robot is able to be fixed to the wheelchair and helps the disabled person in his/her life domestic tasks. At the same time the robot is able to work in the day-life environment by climb the walls, tables and other surfaces. This operation is performed by using the special designed docking mechanisms in the robot and docking stations (DS) located in the home area (rooms, kitchen, bath-room, etc.). The location of the DSs permit the robot to moves from one location of the environment to another, and sometimes from one room to another.

To move the robot from/to the wheelchair to/from the home area and the corresponding DSs, the MATS robot is able to "jump", i.e. make a transition from/to the wheelchair environment to/from DS located in the room environment. Fig. 1 shows MATS robot working in assistive application in the lab environment. (G. Bolmsjo, 2002).

Fig. 1. MATS arm during food task

User requirements will drive the research and development processes in the project with continual user evaluation and peer review of the results obtained at every stage. Every relevant aspect of the lives and environments of potential users will be explored in detail by acknowledged experts in their field. Physicians, therapists and psychologists will contribute to the process of eliciting and evaluating the views and expectations of the end users who can benefit most from the application of the MATS system. The results of this user requirements study will be used to generate functional and performance specifications for the system's components, which will then be designed and manufactured to satisfy the users needs. A great deal of functional flexibility and versatility will be derived from the use of software and the integration of the system into 'smart home' environments.

2. MATS ROBOT DESIGN

The MATS robot final design has five degrees of freedom, and it divided in two parts: the tips that have a docking mechanism to connect the robot to the wall, or a wheelchair and a gripper. The body has two links that contain the electronic equipment and the control unit of the arm. It is important to note that the robot is symmetric (Virk, 1999), due to it is possible to fix the arm in any of its ends. The raw material is made by aluminium and carbon fiber. The actuators are torque DC motors, and the used gears are Harmonic-Drive. The range and the position of the different joints can be seen in figure 2. Figure 3 shows the real prototype.

Fig. 2. MATS mechanical drawing

The power supply is taken from the connector that is placed in the center of the docking station in the wall. All the electromechanical and electronic equipment are inside the arm. The robot has all the electric motors and gearboxes and the electronic equipment on-board: amplifiers, encoders, the axis control board, and the main CPU in order to communicate with the user of the arm. The design is very similar to a ROMA climbing robot (C. Balaguer, 2002).

3. MATS ROBOT SOFTWARE ARCHITECTURE

A good user interface is necessary for the acceptance of service robots in rehabilitation, it will be only effective if the underlying system has a certain degree of intelligence (K. Kawamura, 1995). For these reason all the information and the algorithms related to the good working order of the whole system. There are three different computers:

- Main robot computer, the Arm Controller AC

Fig. 3. MATS system

- User PDA (with the Human-Machine-Interface HMI)
- Room computer RC

The software modules are shared between these three computers. The RC contains all the information related to the environment and the different programs that can be used for the robot.

An important role in the MATS-system is played by the HMI, the device available to the user: to command the arm functionality, to be informed about the state of the device or the task the arm is involved, to benefit from the HMI navigational feedback during the transfer maneuvers, to get access to standard application software, including Internet browser and e-mailer. In regard to the navigational feedback, remind the option that the MATS-arm is transferred from (some) permanently installed work-sites to the wheelchair or vice versa. A prerequisite for the implementation of these transfer maneuvers is that precise geometrical relationships are established between the "fixed" socket and the "mobile" socket. The HMI is connected to both the AC and the RC by means of a wireless link, based on the IEEE 802.11b standard as figure 4 shows, and figure 5 shows the global scheme of this software.

4. SYSTEM SOFTWARE ARCHITECTURE

The operator transmits the different commands using the PDA that sends the information to the control unit of the robot. It is better to use a wireless communication in order to minimize the number of wires in the different docking stations of the room. The data to transmit between the two computers are very easy and the different commands are very short in order to reduce the communication time.

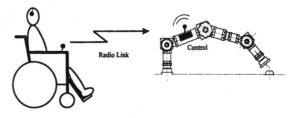

Fig. 4. MATS communication

In the computer room will be implemented the high level tasks. It will exist a database with a library of different automatic programs which includes the different autonomous tasks in the kitchen, in the bath-room, or in different rooms of the home.

This computer has a learning module in order to learn the different movements that the user can make for different daily tasks. In this way, it would be possible that the robot can learn to move to specific places of the environment, and learns to move automatically from one DS to another.

The user-oriented HMI can be used into two different modes: pre-programmed movements and user-controlled robot movements. During motions in the former mode, only objects (e.g. dishes) placed on a tray at fixed positions are manipulated; in the latter mode, the complete control is delegated to the user and allows movements in the entire workspace of the robot arm, either in the joint mode or the Cartesian mode. For the sake of user's safety, we also assume that the user is always in the control loop, at least with the role of supervisor, and can override the current control actions in the case that she/he is not satisfied by the system behavior.

It will be possible to include more pre-programmed task developed in the future, even by other companies. This task could be incorporated to the PDA or the Room Controller in different ways (i.e.: via internet).

Currently, the computer for the MATS-HMI prototype is an handheld Personal Computer (PC) (Mod. iPAQ H3870 manufactured by Compaq Computer Corporation, Houston, TX, U.S.A.), running the MS Windows CE 3.0 operating system (OS). Three ports are available for interfacing to outer devices: a wireless port, that requires an external radio card (Mod. COMPAQ WL110), a serial communication interface, or an infrared port at 115 Kbps, based on the IrDA standard protocol of communication (currently unused).

Inside the robot, there are a board (manufactured by Intrinsyc). Its main characteristics are a CPU at 192 MHz., 3 serial ports (used to connect with the motor's drivers), USB port, Ethernet connector and compact flash slot (used for the

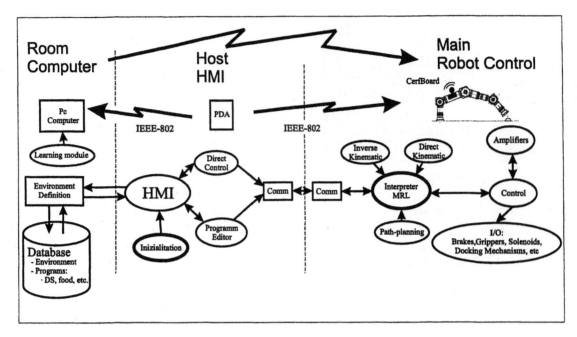

Fig. 5. System software architecture

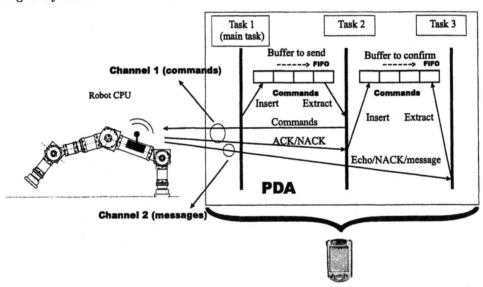

Fig. 6. Communication protocol

wireless card), running the MS Windows CE 3.0 operating system (OS).

When the robot program starts up, waits indefinitely for connections from any client. When the program in the PDA starts up, tries to connect to the robot. Then the connection is made and they can begin to interact. In that moment the commands flow can start and the robot could begin moving and do tasks. Figure 6 shows the communication protocol and dataflow inside the PDA.

The software elements of the MATS, HMI and robot's application, are written in MS eMbedded Visual C++ 3.0. From a functional viewpoint, the program can be subdivided into three main parts. The first part of the program is intended to invoke the specific service requested by the user and interface control (screens, button). The second part is focused in the communications protocol. It is necessary to let the user interact with the interface in any moment at the same time that the application is exchanging information with the robot. The last part is in charge of voice recognition, this is another way to control the application explained in the next paragraph.

For this purposes, the applications use multitasking in order to allow all this task run simultaneously.

There are two ways to interact with the HMI. Like a normal application, clicking buttons, writing text and values using the stick or even the finger because the screen is tactile, see figure 7. Another

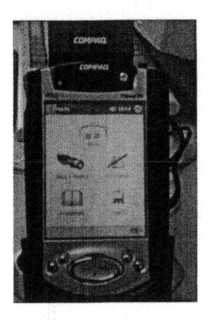

Fig. 7. HMI PDA

Fig. 8. HMI Teleoperation PDA

way, in order to allow the use of this robot to the most severely disabled people, is via voice. The application has a voice recognition module. The user can use a set of defined commands to manage the interface and can perform exactly the same tasks permitted using the tactile screen. Also, the HMI has a voice synthesis module to generate voice message to the user when he/she is controlling the robot using the voice.

4.1 Environment

There are three different kind of DS:

- Fixed DS. This kind of mechanisms are fixed to the walls and the others places of the house where it is needed for any special task such

as in the table for placing the plates into the dish-washer.
- Mobile DS. When the robot needs to move a large distance between two DS it is better to move in high velocity. This is possible if the DS can move in a rail into the wall.
- DS inside the wheelchair. It is a special DS, where is located inside the wheelchair. There is a special DS in the room where allow the transition between the room DS and wheelchair.

4.2 MATS robot language (MRL)

An own language and communication protocol has been developed to implement the interaction between the PDA and the robot. Such language has been called MRL (MATS robot language).

The PDA sends to the robot's CPU, and vice versa, a set of commands pertinent to the last MRL language. Some examples of this commands includes: $MOVC\ p1,p2,p3,p4,p5$, $MOVP\ p1,p2,p3,p4,p5$, where $p1,p2,p3,p4,p5$ are numerical parameters which describe the end position of the free tip of the robot, using the screen that in figure 8 can be seen. Another commands in this language are related to handle the digital inputs and outputs such as GROP, GRCL (to open and close the grippers. A communication protocol has been specifically developed to define the commands and messages exchange mechanism between the HMI and the robot (ACKs, NACKs, echoes).

When a command movement is received by the on-board control program, they are first parsed and introduced in a queue waiting to be processed, in the way the robot can be always listening for new commands at the same time others are being processed. Then the command is executed and the corresponding motion primitive are calculated using the robot internal inverse and direct kinematics model as required. The robot uses another kinematics equations to calculate the trajectory in a straight line, one of the most complex movements because require lots of computes and a high degree of precision to do an accurate movement.

The CPU inside the robot is equipped with the facility of a low level programming to issue direct commands to the motor's drivers. The latter is provided by the manufacturer with its own low level programming language. A direct communication with the controllers serves for fault diagnosis, resetting, calibration, changing control parameters and assist in performing other important maintenance functions such as enquiring about some states of motors and the current control gains for each of them.

5. SOFTWARE IN THE ROBOT

The robot has an embedded PC. This CPU is an $Intel^{TM} StrongARM^{TM} SA - 1110$ (velocity 190 MHz), which has to implement the direct and inverse kinematics of the robot. We need to implement the MRL interpreter to understand which is the commands that the user send via radio ethernet.

The main tasks in this computer are:

- Communication protocol (to interact with the client, PDA)
- Commands interpreter
- Kinematics transformations (Direct kinematics and Inverse kinematics)
- Path-planning (for straight line movements)
- Connection to the amplifiers
- Commands to digital inputs and outputs (i.e.: open and close the grippers)

For robot security it is impossible to release one docking mechanism until the other one is safely locked to the docking station. Teleoperation commands are not executed until previous commands are terminated or aborted. The new joint values and other data are sent regularly to the ground computer and the information is updated on the Teleoperation Interface.

A variety of software and hardware fault reporting and recovery are implemented during the design and development of the robot. These are issued to the user in form of warnings or error messages, containing hints or solutions suggestions.

6. SUMMARY AND CONCLUSIONS

The MATS robot software architecture and its HMI is modular. The communication of the robot control unit with the main computers is wireless via IEEE 802.11b protocol. The received and sent commands corresponding to the developed MRL (MATS robot language) standard are very easy to learn. The system will be able to adapt in an unstructured environment, and it is possible assist more complex tasks. MATS is designed to be modular and capable of fitting into any environment. This means that for the first time a robot can move accurately and reliably between rooms and up or down stairs, and can transfer from being wheelchair-mounted to floor, or wall-mounted. This degree of flexibility will have significant implications for the care of the disabled and elderly people with special needs. The modularity of the system is able to grow as the same level of disability of the user changes.

MATS sockets in an office would also be used for offering structured and unstructured assistance with daily living activities such as drinking, eating, picking items up etc.

The MATS robot, developed by the University Carlos III of Madrid, is under lab testing period. The robot has excellent weight/arm-length ratio taking in mind that all the control hardware is on-board.

ACKNOLEWDGEMENTS

This work has been funded by the EU community under project IST-2001-32080. The authors would like to acknowledge the work of the other partners involved in this EU project.

REFERENCES

C. Balaguer, A. Gimenez, M. Abderrahim (2002). Climbing robots for inspection applications of steel based infrastructures. *Industrial Robot, An international journal* **29**, 246–251.

G. Bolmsjo, H. Neverid, H. Eftring (1995). Robotics in rehabilitation. *IEEE Transactions on rehabilitation engineering* **3**, 77–83.

G. Bolmsjo, M. Olsson, U. Lorentzon (2002). Development of a general purpose robot arm for use by disabled and elderly at home. In: *33rd International Symposium on Robotics (ISR)*.

K. Kawamura, S. Bagchi, M. Iskarous M. Bishay (1995). Intelligent robotic systems in service of the disabled. *IEEE Transactions on rehabilitation engineering*.

K. Wada, T. Shibata, T. Saito K. Tanie (2002). Analysis of factors that bring mental effects to elderly people in robot assisted activity. In: *International Conference on Intelligent Robot and Systems (IEEE/RSJ)*.

Kwee, H. (1997). Integrated control of manus and wheelchair. In: *International Conference on Rehabilitation Robotics (ICORR'97)*.

L. Zollo, C. Laschi, G. Teti B. Siciliano P. Dario (2001). Functional compliance in the control of a personal robot. In: *International Conference on Intelligent Robot and Systems (IEEE/RSJ)*.

Topping, M. (2002). An overview of the development of handy 1, a rehabilitation robot to assist the severely disabled. *Journal of intelligent and robotic systems* **34**, 253–263.

Virk, G.S. (1999). Technical task 1: Modularity for clawar machines - specifications and possible solutions. In: *2nd International Workshop and Conference on Climbing & Walking Robots (CLAWAR'99)*.

Copyright © IFAC Telematics Applications in Automation and Robotics, Espoo, Finland, 2004

ELSEVIER
IFAC
PUBLICATIONS
www.elsevier.com/locate/ifac

REMOTE OPERATION OF THE RADIATION-PROOF ROBOT USING THE DIRECT MASTER ARM SYSTEM

Kiyoshi Oka[1], Hiroki Miura[2] and Goro Obinata[2]

[1] *Japan Atomic Energy Research Institute, Tokai-mura, Naka-gun, Ibaraki, JAPAN*
[2] *Nagoya University, Furo-cho, Chikusa-ku, Nagoya-City, Aichi, JAPAN*

Abstract: In Japan Atomic Energy Research Institute, we have developed a radiation-proof robot (called RaBOT) for accomplishing a given mission under severe conditions such as gamma radiation. RaBOT can be maneuvered by remote commands from the operation desk with the master arm system. However, the system has become large and complicated because the master arms have the same number of active joints as the slave manipulators; moreover, that provide not only the required sensory feedbacks on the operator but also uncomfortable constraints. This report examines the performance of teleoperation for the master arm system and proposes a new type of master arm as the alternative. *Copyright © 2004 IFAC*

Keywords: Master-slave systems, Nuclear plants, Remote control, Robot, Teleoperation, Telerobotics, Training

1. INTRODUCTION

A critical accident at the JCO occurred in September 1999, which was the most serious nuclear accident in Japan (Tanaka, 2002). It took long time to collect the necessary information on the accident and then to shut down the critical reaction since human access was limited due to high radiation level. According to such the background, the Japan Atomic Energy Research Institute (JAERI) has developed a radiation-proof robot (called RaBOT) (Oka and Shibanuma, 2002) which can operate under high γ-radiation field up to 10^5 Gy. The RaBOT is a mobile type robot equipped with dual articulated manipulators, four crawlers, CCD cameras and so on. In accident fields, the man-machine interface, which transmits information to the operator efficiently, is required in order to operate RaBOT remotely. A seat-type control desk is adopted to RaBOT, and it can be operated during a certain time through cable or wireless communication line from the control desk in a movable container. The control desk is equipped with the master manipulator to maneuver the dual articulated manipulators of RaBOT. Bilateral control function is installed in the master arm and the slave manipulator; that is, the force at the end-effector is reflected to the operator while RaBOT touches object. The force reflection to the operator is important and required so that the operator can feel remotely the realistic sensations of the teleoperated robot. Then, the operator can carry out various tasks by using the man-machine interface remotely. However, the master arm including the power transmission mechanism causes the enlargement and complication of the operation system. Moreover, not only pressure and restraints are given on the operator arms and hands but also the carriage of system to the accident fields becomes difficult.

In this paper, we study the improvement of the portability and the remote controllability, and we try a new method in which the human's arm works as the master-arm of the teleoperation for simplifying the existent system. In addition, the operator without training can operate the RaBOT manipulator with the proposed system. Then, this paper describes the outline of RaBOT and the improvement of the human interface system of RaBOT.

2. OUTLINE OF RaBOT

Environmental conditions are listed in Table 1 for the development of the RaBOT. In particular, the dose-rate of the γ- radiation is higher by one order of magnitude than that of the critical accident at the JCO. Based on the conditions, RaBOT has been designed and fabricated. For application in a radiation field, RaBOT is composed of radiation-resistant parts for the main mechanisms, such as the manipulator, crawler, battery and power supply, except for the electronics parts with low radiation resistance. Electronics parts such as CCD camera and control device are modularized for remote replacement in a short period so as to continue a rescue mission in the accident facility.

Table 1 Environmental conditions

Item	Conditions
Gamma radiation	~10Gy/h (Total dose : ~10^5Gy)
Atmosphere	Air, radioactive dust
Temperature	~40 degrees C
Humidity	~90% (no dew condensation)
Traveling area	Standard passage for inspection made of concrete, asphalt, linoleum and grating in the facility
Stair	90 cm in width, 22 cm in step height, 40 degrees in inclination angle and 120 cm in landing width
Door	90 cm in width and 180 cm in height after opening the door

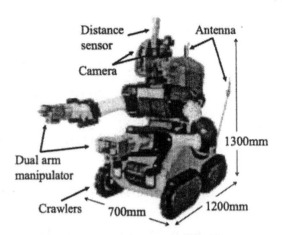

Fig.1 Radiation-proof Robot (RaBOT)

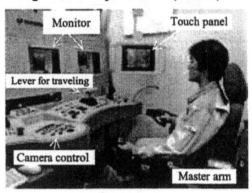

Fig.2 Control desk

Figure 1 shows an overview of the fabricated RaBOT. The RaBOT is 700 mm in width, 1200 mm in length and 1300 mm in height, specified by the required passage of door and stairs. The manipulator is equipped with a force sensor at the wrist, so that force-feedback control of the manipulator is applicable to dexterous operation. Four crawlers with independent torque control as traveling mechanism are adopted for the stable passage through stairs with 40 degrees, and travel with a speed of 2 km/h on the flat floor. The RaBOT is operated by wire or wireless control from a control desk. Eleven removable modules with low radiation resistance such as servo amplifiers, CCD cameras and control devices are installed on RaBOT, and these modules are easily replaced by human or another robot.

The RaBOT is controlled by one operator from a control desk in a movable container, as showed in Fig.2. The movable container is 2.4 m in width, 6 m in length and 2.6 m in height. Two levers or two foot-pedals control the respective traveling speed of the four crawlers. The operator can also operate the dual manipulators of RaBOT by using two master arms, which are the armrest type 6 axes joystick with force-feedback control. In addition, the control desk is equipped with a camera control panel for changing the image information from RaBOT and a touch panel for changing the position or force control mode of each axis of the manipulators. In particular, for force control mode, the operator can handle the manipulators through the master arms after selection of unilateral or bilateral mode. In addition, a compliance control can be applied to the manipulators by selection of suitable parameters in the force feedback control.

3. DIRECT MASTER ARM SYSTEM

The operator can carry out the various tasks remotely from the control desk in the movable container. However, it is difficult that the operator feels the force feedback exactly. The skill is required to operate the dual manipulators of RaBOT smoothly. In addition, it is not easy to carry the control system to the accident field because of the large size and the heavy weight. According to such the background, we study the improvement of the carriage and the remote controllability, and we try a new method in which the human's arm works as the master arm for the purpose of miniaturization and simplification of the existent system.

3.1 New concept of the master arm

We have studied the direct master arm system (Iwami, *et al.*, 2001) in order to minimize the existent master arm system and operate the manipulator of RaBOT easily. The proposed new master arm system and RaBOT are shown in Fig.3. There are remarkable features in the teleoperator. First, there is no mechanical part in the master arm system. Three sets of small electrodes and two position / orientation sensors are attached on skin of upper limb. The upper limb works by itself as the master arm. Second, force reaction from the manipulator of RaBOT to the master arm is achieved by using functional electric stimulation (FES) through the surface electrodes. Third, the structure of the manipulator of the RaBOT is kinematically similar to the master arm. This makes it easy for the operator to command RaBOT. In other words, the operator can operate RaBOT manipulator without training. The master arm system of this teleoperator

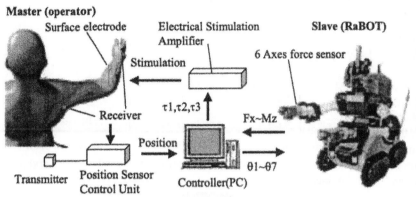

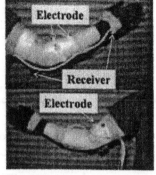

Fig.3 Direct master arm system with FES   Fig.4 Master arm (operator)

is shown in Fig.4. Two magnetic position sensors on the operator arm measure the position and orientation in the master coordinate frame. The master arm can be modeled by a seven degree-of-freedom rigid link mechanism. So, we can calculate in real time the corresponding angles of the arm joints from the measurements, and then the calculated angles are sent as position commands to RaBOT. The force sensor, which is placed at the wrist of the manipulator of RaBOT, measures the generated force and moments. Those are converted to the corresponding torques of the joints, which are reflected to the master for increasing the sense of telepresence of the operator. The reaction is conducted by stimulating the corresponding muscles. The operator feels the artificially generated forces as a reaction force with visual information. In comparison with the existent master arm system, the operator with this new master arm system manipulates more freely the manipulator of the RaBOT because there is no physical constraint on his arm and hand. In addition, the carriage of the operation system is improved.

3.2 Force feedback by FES

The FES is a therapy, which reconstructs function for paralyzed muscle by giving electric stimulation. A juncture of muscle and nerve is called the motor point. When rectangular current pulses get to the motor point through the nerve, the muscle contracts. When FES is utilized for rehabilitation medical treatment, rectangular current pulses are given to the motor point from the outside of the skin, and the frequency of impulse current is 20[Hz]. By using this technique, we can control the muscles force from outside the body.

The force information is detected by a tendon as muscle tension. When there is a difference in the tension of the antagonist, a human feels it like as if external force acts on his arm. Thus, we can artificially transmit force information with the stimulation to the antagonist of the operator's upper limb. If the bend muscle is stimulated by the FES,

the operator increases the muscle force of the extension muscle in order to maintain the attitude. In this case, the tension of bend muscle and extension muscle becomes equal. Therefore, it should not be sensed that an external force was received from information of tendon. However, as a result of the experiment, many examinees felt this stimulation was an external force. Because the brain must send stronger order for the extension muscle in order to maintain the attitude. The brain judges it is the result of receiving the external force. Using this phantom sensation, it is possible to transmit the force reaction to the arm of the operator by the FES. In the experiment, each muscles of dorsiflexion direction (wrist joint 1) and ulnar flexion direction (wrist joint 2) and elbow joint are stimulated as shown in Fig.5. The force reaction of these joints is necessary in order to know the direction of the force which the slave end-effector received. And the motor points of these muscles are easy to be stimulated using the surface electrode, since there are near the skin. The muscle forces are controlled by amplitude voltage of stimulation. Here, when the frequency of the stimulation is higher, the fatigue of the muscle quickens. So we chose 100[Hz] as the frequency of stimulation in the experiment.

4. EXPERIMENT WITH DIRECT MASTER ARM SYSTEM

In order to confirm the effectiveness to a new operator, we have carried out two experiments using the system shown in Fig.3.

4.1 Slave arm operation

The first experiment is a basic manipulator control. Figure 6 shows the appearance of RaBOT manipulator operation by the direct master arm system. Two magnetic position sensors and three electrodes are set on the operator arm. Between RaBOT and the controller (personal computer) are connected by wired Ethernet in order to simulate the delay of the operation by wireless method. In this experiment, the control period is fixed to 100msec, because the control period is unstable by reason of Ethernet connection. Figure 7 shows three-dimensional trajectories of the circular movement of the manipulator tip. As the result, it is confirm that RaBOT manipulator is moved smoothly along the operator action, although a little time delay is included. The direct master arm system is adopted, so that the operator can move RaBOT manipulator without considering the singular point. From this experiment result, it is clarified that the unpracticed operator can easily operate RaBOT manipulator.

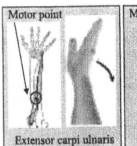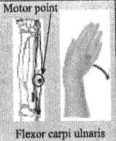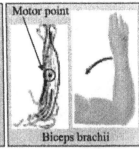

Motor point	Motor point	Motor point
Extensor carpi ulnaris	Flexor carpi ulnaris	Biceps brachii
(a) Wrist Joint 1	(b) Wrist Joint 2	(c) Elbow Joint

Fig.5 Forearm motions controlled by FES, stimulated muscles and motor points

4.2 Reaction force by FES

The second experiment is the confirmation of the reaction force by FES. The RaBOT manipulator is extended straight in front as shown in Fig.8. The RaBOT manipulator is pushed in every direction and the reaction force occurred on the manipulator is measured. Figure 9 shows the experiment result of the FES. As the result, in the case of the manipulator is pushed upward, the value of +Fx direction increased, and the amount of stimulation of the biceps brachii was in proportion to the amount of increasing of the manipulator movement. In the same way, when the manipulator is pushed to the +Fz and −Fx direction, the muscles of dorsiflexion and ulnar flexion of the operator's wrist joint are stimulated. The operator can catch the amplitude and direction of the generated force at the manipulator end-effector by this force reaction. However, the arm of the operator didn't move inside though force was increased in the -Fz direction. It is because there is no electrode to bend the wrist of the operator inside.

5. COMPARISON TEST

The same work was carried out to compare the performance of the new master arm system with the existent master arm system. As a common task often, a door-opening task is selected. In order to open a door, the manipulator grasps the doorknob and rotates it with only a certain fixed angle, then, the latch bolt is stored. The door is opened along the rotation of the door. When the doorknob is grasped, the reaction force is occurred between the end-effecter and the doorknob. The reaction force is informed to the operator through the force sensor of the manipulator. The existent master arm system is set to the bilateral control mode. The rate of force to position is set 1.

Figures 10 and 11 show the result of each operation method. Fig.10 shows the change of each

Fig.6 Remote operation test by the direct master arm system

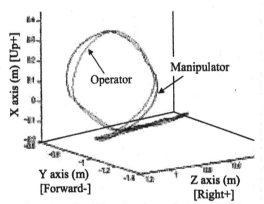

Fig.7 Result of the circular trajectory (3-D)

joint angle by the existent master arm system. All joint angles vary smoothly and the door is opened within a short duration. On the other hand, new master arm system needs the long time to open the door, and all joint values are oscillated as shown in Fig.11. This is caused by in sampling time's being slow. Figures 12 and 13 show the calculated torque to stimulate each muscle and the amount of the stimulation to give to it to the operator, respectively. It is understood that the amount of stimulation is always maximum value. Smooth door-opening work couldn't be done though the performance improved a little by doing the adjustment of the amount of the reaction force. From this result, it can't be said that it is still utility about the direct master arm system with the stimulation by FES. However, as for the operation of RaBOT manipulator, training was not so necessary to the operator, and it was found out that work by RaBOT manipulator could be easily done comparatively. In addition, it is possible that work is carried out without operator worrying about the movement in the neighborhood of the singular point which is characteristic of the different structure type master arm system. As for that point, the improvement of the operation could be confirmed.

6. CONCLUSION

We have developed a new master arm system with FES force reaction for the radiation-proof robot (RaBOT) so as to solve control problems of teleoperation and portability. It is shown in experimental results that the developed master arm system with the designed controller is more effective to cope with phase transition problem and to increase sense of telepresence in comparison with existent master arm system.
We obtain the following results.

(1) A direct master arm system was adopted, and it turned in the miniature and light weight the existent master arm system.
(2) An operator can operate the RaBOT manipulator without training. In other words, it is training free.
(3) It is possible that work is carried out without

operator's worrying about the movement in the neighborhood of the singular point which is characteristic of the different structure type master arm system.
(4) It was confirmed that this method can artificially transmit force information with the stimulation to an operator. In other words, the operator can feel the reaction force from the manipulator.
(5) A comparative experiment with the existent master arm system was done, and the availability of the direct master arm system was shown.

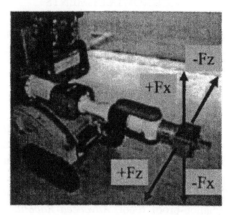

Fig.8 Reaction force test with FES

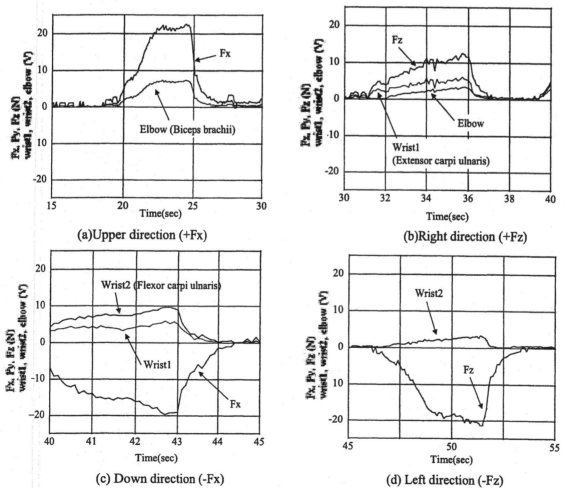

(a)Upper direction (+Fx)

(b)Right direction (+Fz)

(c) Down direction (-Fx)

(d) Left direction (-Fz)

Fig. 9 Reaction force on the direct master arm system

(6) It couldn't be operated by the actual work normally due to the delay of the sampling time (Anderson and Spong, 1989). However, it confirmed that the improvement of the operation could be expected by the improvement of the control technique (Kosuge, *et al.*, 1996).

After this, a remote control test will be done, and working efficiency by the remote work and the improvement of the operation will be examined. In addition, the improvement of man-machine interface will be done more.

ACKNOWLEDGMENTS

The authors would like to express their sincere appreciation to Drs. M. Seki and H. Takeuchi for their continuous encouragement on this work.

REFERENCES

S. Tanaka (2002). Summary of the JCO criticality accident in Tokai-mura and dose assessment, *Journal of Radiation Research*, **42(Suppl.)**, pp.S1-S9.

K. Oka, K. Shibanuma (2002). Development of a radiation-proof robot, *Advanced Robotics*, **16-6**, pp.493-496.

T. Iwami, G. Obinata, A. Nakayama, H. Miura (2001). Force Reflection by Functional Electrical Stimulation in Teleoperator, *Proc. of the SICE/ICASE Joint Workshop*, pp.205-210.

R. J. Anderson, M. W. Spong (1989). Bilateral Control of Teleoperators with Time Delay, *IEEE Transactions on Automatic Control*, **34-5**, pp.494-501.

K. Kosuge, H. Murayama, K. Takeo (1996). Bilateral Feedback Control of Telemanipulators Via Computer Network, *Proc. of the 1996IEEE/RSJ Int'l Conf. on Intelligent Robots and Systems*, pp.1380-1385.

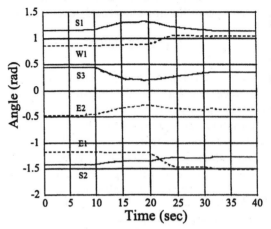

Fig.10 Door opening test by the exist master arm system

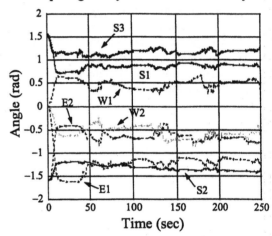

Fig.11 Door opening test by the direct master arm system with FES

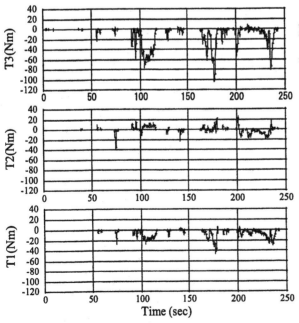

Fig.12 Calculated torque for the direct master arm system

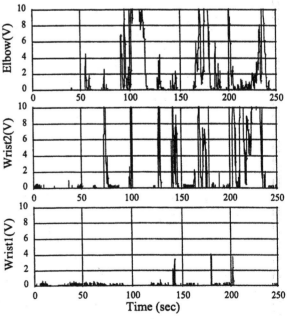

Fig.13 Stimulation quantity to the operator corresponding with the calculated torque

Copyright © IFAC Telematics Applications in Automation and Robotics, Espoo, Finland, 2004

ELSEVIER
IFAC
PUBLICATIONS
www.elsevier.com/locate/ifac

HARDWARE AND SOFTWARE ARCHITECTURE FOR RAPID AND SAFETY USE OF ROBOT SYSTEMS

Mikko Sallinen[1], Sakari Pieskä[2]

[1]VTT Electronics, Kaitovayla 1, 90571 Oulu, Finland
[2]CENTRIA Research & Development, Vierimaantie 7, 84100 Ylivieska, Finland

Abstract: In this paper, we present a hardware and software architecture for industrial robot system where rapid ramp-up and safety tele-control is considered. The architecture is designed to be open and portable to a different hardware platforms and applications by fulfilling the requirements of time and change of information. The key issues are easy tele-operation and maintenance while system failures will occur. In the experimental phase, we illustrate few examples where the developed system will be tested. *Copyright © 2004 IFAC*

Keywords: software architecture, remote use, robot workcell

1. INTRODUCTION

In the flexible production, time used for changing the tools, jigs and settings is increasing all the time. Size of lots manufactured is decreasing when production of large lot sizes is moving to other units of the companies. Life cycle of products is decreasing and even the whole manufacturing lines are repositioned in the factories. This sets requirements for the machines operating in the manufacturing. They have to be fast to install and start the production when relocating the physical line and flexible when changing the production for different products.

Research on software architectures in robot applications focuses on task level control of mobile robots (Camargo *et al.*, 1992) and especially for teams by Parker (Parker 1998). There are also examples on software architecture on FMS cell, but they lack on information considering safety use and comfortable recovering from the failure of the production (Weiss & Konieczny, 2002). An overview of research carried out in higher level is presented by (Hamilton *et al*, 1994) and an example from fault tolerant system on industrial systems can be found from (Tyrrel & Sillitoe 1991). There are also some examples of task level control systems over internet without considering safety properties (Ghaffari *et al*, 2003).

This paper considers a software architecture describing the overall software architecture and details of it in chapter 2. Chapter 3 illustrates the communication between the layers and after that

examples where the architecture is going to be applied are illustrated in chapter 4. Conclusions of the system are presented in chapter 5.

2. DESCRIPTION OF THE ARCHITECTURE

Approach to the architecture is a general purpose, easy-to-generate, easy-to-use and effective is necessary. Advantages of use of the general purpose are same structure in all the machines in the production, which simplifies the overall control of the system. General purpose is described here including open architecture for both hardware and software platforms. Easy-to-generate means that most of the code can be generated automatically. The software is divided into several layers and modules, which can be generated separately, and the goal is that these software modules can be generated automatically using generation tool. Levels of communication between the software modules in the system describe the effectiveness of the communication system. System has to be capable of managing tasks from hard real time to open time-limit tasks.

2.1 General description

The system is divided into two main parts: general part and device dependent part. General part is aimed to be similar for all the same-functioning machines when device dependent part is specified for each machine separately. The architecture is described in figure 1:

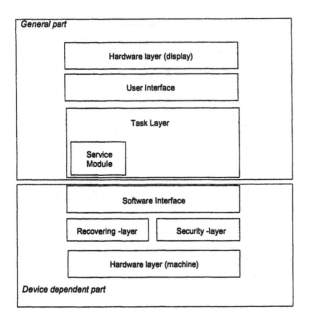

Fig. 1. Architecture for teleoperation

The dotted lines separates the two main parts of the system that are further divided into several subparts.

2.2 General part

2.2.1 Hardware layer

Hardware layer illustrates the information to user using display, which can be mobile or fixed. Examples can be standard monitor, handheld PDA or mobile phone depending on the requirements.

2.2.2 User Interface

When data to be illustrated to user or other machine is obtained in different formats, it has to be processed into simpler form. User interface carries out this task whether the time requirements for illustrating the data is fulfilled.

The user interface can be carried out in standard graphical application, coded in HTML or XML. For applications where a lot of remote use is needed, the web browser –based user interface is recommend.

In the case of mobile user interface it can be illustrated in mobile phone, PDA or on industrial PC designed for mobile machines. In all of these solutions, interface may be text or graphical –based.

2.2.3 Task layer

Task layer contains the main application. Physically task layer may be distributed for several devices or machines. In the case of distributed system, these different parts communicate with each other using secure communication lines.

The advantages of distribution are possibilities for using different programming languages and storing the larger data to most suitable platform. Computer – controlled industrial machines usually does not provide large storage space for data and in those cased it is better to store data into external storage. Task layer contains often real-time applications and in those cases distribution is the only way to run the applications where non-real-time parts and real-time parts are included.

Inside the task layer is a service module, which contains the data and rules for communication for Recovery and Security –layers. It also may contain a program for carrying out tasks required for Recovery and Security –layers.

In the case of several external devises in the system, each has it's own task layer and device dependent part under it. The hierarchical description is illustrated in figure 2.

2.3 Device dependent part

2.3.1 Software interface

Main purpose of the software interface is to take care of communication between programs in task layer and programs in Recovery and Security –layers. This is possible by device –dependent drivers that are locating in software interface. If there are several devices, each of them has an own software interface.

2.3.2 Recovering layer

There are two main tasks of Recovering layer: report on errors and failures during the execution if they are managed by own recovering system or in the case of serious failure when assistance of external workforce e.g. user is needed, help and assist user for repairing the system. Programs of Recovering layer may use data stored in Main or Sub Task layer, but the master control is located in Recovering layer.

Reporting of the errors can be logging the error data into file or display wherefrom the user can follow the current state of the system. In the case of serious failure, Recovering layer tries to locate the source of error by observing the status of the different parts of the system and history of logging the error data.

Recovering of the failure can be divided into two main models: fully autonomous and semi autonomous. In the fully autonomous case, system can recover from the failure by itself while in semi

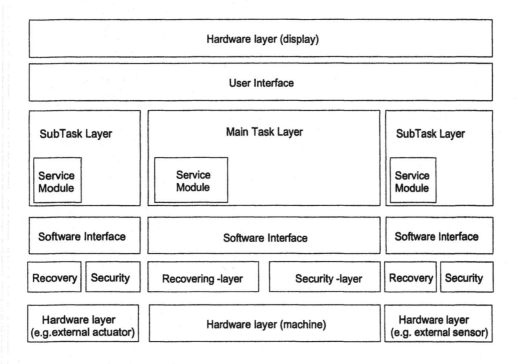

Fig. 2. Extended architecture: several devices in the system

autonomous model by giving the instructions to the user, system can be recovered. User here means worker of a machine, not an expert of repairing systems or machines. Autonomous cases can be divided as follows:

1) System recovers without stopping the process
2) Process is halted but system can recover by itself
3) User have to assist in the recovering process

The similar structure of alternatives of recovering while user assist the system can be written:

1) User can recover the system without halting the process
2) User can recover the system but the process is halted
3) A special assistance is needed e.g. professional help over the web

For assisting the user, Recovering layer uses internal and external sensors of the system for observing status. In many cases, system is halted and information only from external sensors is available. This is possible only if external sensors are separated from the main task.

2.3.3 Security layer

Purpose of the Security layer is to take care of security of the system including both securing the

context of data transfer and time requirements of transferring the data. In addition to that, security layer controls the access of the users for using the system. Identification can be carried out by password, fingerprints or iris of the eye. Service module in Task layer takes care of storing the data for Security layer.

2.3.4 Hardware layer

Hardware of the system is located in hardware layer. The main hardware is located under main task layer and possible external devices have their own structure including Recovering and Security layers.

3 COMMUNICATION BETWEEN THE LAYERS

To fulfill the requirements of reliability, usability and operation, there is a need for describing the rules for transferring the data in the system in a terms of type of data, when and how it will be transferred.

3.1 Methods of transferring the data

Information is transferred in horizontal and vertical direction. Vertical transformation is carried out between each intersection layer but horizontal communication is only between Task layers. Reason for limited horizontal transformation is modular structure in which each device presents own unit with own control. The advantage of this structure is that devices can be changed flexible without complicated configuring and rebuilding tasks. In this model, only the Task layers communicate with each other.

Directions of communication can also be named as internal communication in the case of vertical communication and external communication in the case of horizontal communication. Internal communication is focused on effective data transfer while point in external communication lies on safe transfer of information.

3.2 Time requirements

There are time requirements for operation and communication between each layer. The requirements in the architecture can be divided as follows:

1) Hard real-time operation
2) Fast operation
3) Not time-dependent operation

The highest requirements for operation and communication is set on Hard real-time operation. In that kind of layers operation has to be designed certain way from the very beginning of the system to fulfill the requirements. Operation is based on fixed control loops of execution of tasks and the time space for operation is between 1 ms to 100 ms.

Fast operation is not such a strict about time than hard real-time operation but operation have to still be run fast. Time space for operation is between 100ms to 2s. Not time-dependent operation and communication is everything slower than fast operation. It does not have any fixed requirements for carrying out the task.

3.3 Interfaces for communication

Interfaces for communication can be divided based on types of transformation protocol. There are several possibilities for general part communication in which nowadays TCP/IP protocol is highly encouraged. Device –dependent part is restricted to physical communication lines of each machine. However, nowadays there are available several possibilities including ethernet, CAN or serial line. Also digital and analog communication can be user for fast communication without versatile change of information.

3.4 Generation of the software layers

To obtain effective results when using the proposed architecture, the generation of the software layers

have be automatized as far as possible. The final application forms the details of the system, but overall description has to fulfill the requirements.

3.4.1 Automatic generation

The goal of the software generation for robotic systems is to generate the software packages automatically. When the software architecture is described with details, generation should be a straightforward task.

The details of each system are determined using parameters which are selected to describe and cover the functions of the different software layers. The actual compiling process can be linked together with common compiler like c/c++ or the code can be generated using special postprocessing tool designed for each specific hardware. Device dependent layers are always generated into format, which is understandable of particular machine.

3.4.2 Manual generation

If the generation of software layers does not contain a set of different machines with small deviations of their properties, software is reasonable set up manually. Advantages of the manual generation is effectiveness and especially is the cases of hard real-time applications where time requirements are tight, manual coding and debugging gives a good solution.

Naturally there are variations between automatic and manual generation where usually some software modules of general part are similar in all the different applications and device dependent part is generated manually. A typical example of this kind of solution can be find is when general software takes care of several types of machines and a device dependent part is coded transparent to the user.

4. EXAMPLES OF THE APPLICATIONS

The architecture was tested in VTT's workcell with industrial robot (ABB IRB1400 with S4 controller). The test case was to sent error messages given by robot to the user with remote interface. This interface can be any computer with internet connection. In the test, the error message was noticed in the Recovery layer. The information provided to the user was stored in Task layer where it was also processed. Information to the user was sent via user interface to the network displayed using web-browser.

The functionality of the proposed architecture will be verified also in additional applications. One of these applications is a robotized material handling cell at Production Engineering laboratory of CENTRIA Research and Development, see Fig. 3. The industrial robot in the cell is ABB IRB 4400 with S4+ controller. The cell is equipped with three different solutions for remote monitoring: wlan-based camera, RF-based observation camera and mobile robot equipped with web-camera.

Fig. 3. Robot cell at CENTRIA.

The industrial verification environments include a robotic welding cell used in flange production at Ferral Components LTD (see Fig. 4) and robotic press brake cells at Celermec LTD and Mecanova LTD which are also the participating companys of this project.

Teleoperation tasks that the system will take care are monitoring of the applications, collecting the production data, safety tele-control in task level and tele-mainteinance.

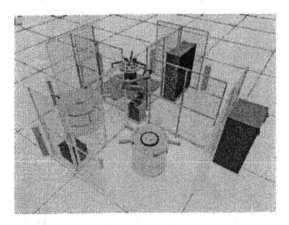

Fig. 4 Robotic welding cell at Ferral Components.

CONCLUSIONS

In this paper, architecture for rapid ramp-up and safety use for robot workcells is presented. The description of the architecture contains hierarchical structure where each layer has it's own task to do. The specified requirements are fulfilled by Recovery and Security layers which takes care of recovery of system or process failure or takes care about the security of system in a terms of data transfer and user identification. An example of the proposed architecture is illustrated in the experimental tests.

Future works include implementation of developed system architecture into industrial use and test it with different robot platforms.

ACKNOWLEDGEMENTS

European Union, National Technology Agency of Finland (Tekes) and four SME companies have financed this work, which is greatly acknowledged by the authors.

REFERENCES

Camargo R. F., Chatila R., Alami R., (1992). Hardware and Software Architecture for Execution Control of an Autonomous Mobile Robot. Proc. Of IEEE Int. Conf. On Industrial Electronics, Control, Instrumentation and Automation, pp. 818-825

Ghaffari M, Sethramasamyraja B, Hall E, (2003). Internet-based control for the intelligent unmanned ground vehicle: Bearcat Cub. Proceedings of Spie Intelligent Robots and Computer Vision XXI: Algorithms, Techniques, and Active Vision, pp. 90-97.

Hamilton D. L., Visinsky M. L., Bennett J. K., Cavallaro J. R., Walker I. D, (1994). Fault Tolerant Algorithms and Architectures for Robotics. Proc of Electrotechnical Conference, pp. 1034-1036.

Parker (1998). ALLIANCE: An Architecture for Fault Tolerant Multirobot Cooperation. IEEE Trans. on Robotics and Automation, Vol. 14, No 2, pp. 220-240.

Tyrrel A. M., Sillitoe I. P. W., (1991). Evaluation of Fault Tolerant Structure for Paraller Systems in Industrial Control. Int. Conf. On Control, pp. 393-398.

Weiss Z., Konieczny R., (2002). Software Structure for a robot operated flexble manufacturing cell. IEEE third International Workshop on Robot Motion and Control, 269-274.

Copyright © IFAC Telematics Applications in Automation and Robotics, Espoo, Finland, 2004

ELSEVIER

IFAC
PUBLICATIONS
www.elsevier.com/locate/ifac

REMOTE MAINTENANCE OF AGRICULTURAL MACHINES

Matti Öhman, Timo Oksanen, Mikko Miettinen and Arto Visala

Automation Technology Laboratory, Helsinki University of Technology
POB 5500, FI-02015 HUT, Finland

Abstract: The goal of Agrix-project is to develop a prototype of an open, generic and configurable automation platform for agricultural machinery. A typical configuration consists of a tractor and one or several implements. Fault tolerance and remote maintenance over mobile networks are essential research topics due to short seasons for agricultural operations especially in Nordic countries. The main purpose of realizing the fast-prototype of the control system in 2003 was to get acquainted with the problems occurring in working with agricultural machines. Experiences from the fast-prototype, some initial tests to realize the mobile communication system and principles of remote maintenance are reported in this paper. The agricultural implement selected for the fast-prototype was a combined seed and fertilizer drill. The tractor was equipped with ISOBUS compatible electronic control unit. A commercial CAN-controller with a high-power digital and analog I/O interface was used as the implement electronic control unit. The architecture of Agrix fast-prototype was designed according to ISOBUS (ISO 11783) standard. *Copyright © 2004 IFAC*

Keywords: Precision farming, automation technology, open systems, configurability, control, telematics, fault diagnosis, human machine interface.

1. INTRODUCTION

In precision farming, cultivation operations, timing and the amount of cultivation material inputs, are adapted to local optimal values relative to the needs of the cultivated plant in accordance with the soil characteristics and nutrient content. The information needed for adaptation comes from laboratory analysis of soil samples taken from planned positions of the field block, other position based measurements and observations of the field, e.g. crop measuring in harvester machine.

Farm information management system (FIMS) is used to manage the position attributed measurement information. The cultivation operations and set point maps for different cultivation inputs are planned with special computer programs in the farm office or in the companies providing precision farming services. To execute the output of these "computer aided cultivation planning systems", position based control systems are needed. A typical agricultural machine system consists of a tractor and an implement. With position based machine control system, the driver can operate and control the machinery during cultivation manoeuvre. The system can automatically control the planned variables, like the feeding rate of seed and fertilizer in drilling, to position dependent planned set points. The goal of Agrix-project is to develop a prototype of an *open, generic* and *configurable* automation platform for agricultural tractor – implement systems. Also *fault tolerance and remote maintenance* are important research topics.

The optimal time period for specific cultivation operations is very short in Nordic countries. The machine control system and the machine itself should be as reliable and operable as possible during this hectic season. If various faults occur, they should be detected, diagnosed and repaired quickly so that that machine can be operational as soon as possible and the cultivation can be continued. Even in the worst case it should be possible to get system into a state in which it can be safely moved from the field to a repair shop. The wearing should be detected before damages, if possible. If these generic machine control systems are used widely enough there will be markets for remote maintenance and repairing services. These services can be provided via the public mobile communication networks, like GSM and later UMTS. The service provider could analyse the faults on-line based on the state information send automatically by the machine control system. After analysing the data the service provider can advice and support the farmer in repairing the machine. It would be even better, if the faults could be prevented with by analysing the early symptoms before the device brakes down. This kind of *fault tolerance, remote fault diagnosis, support* and *remote maintenance* is one of the essential research themes of the Agrix-project.

Open system interconnection means that international communication protocol standards are utilized in order to get the control units of different tractors and agricultural implements from different vendors to communicate with each other. The idea of Open System Interconnection (OSI) comes originally from the ancient ISO standard 7498 ISO (1984). The reference model defined in this standard has been used very widely and it also forms the conceptual basis of the ISO 11783 standard (ISOBUS 2003), which is currently under development.

Generic control system means that it should be possible to use the same automation platform to control different machines, in this case agricultural implements.

Configurability means that control functions can be defined with high-level, usually graphical, tools instead of programming with low level programming language, that are still commonly used in embedded machine control systems. However, in industrial automation applications, high level, graphical configuration tools are widely used and complex system are built from reusable components.

The three-year Agrix-project is introduced at first. The main purpose of realizing the fast-prototype control system in 2003 was to get acquainted with the requirements and problems related to tractor – implement control systems. The experiences gained in implementing this first version, fast-prototype control system, are reported.

2. THE AGRIX-PROJECT

Agricultural environment is challenging for automation. The usage of crop farming machinery is usually seasonal except tractors. Machines are stored inside most of the year and are used only some weeks yearly. The storage conditions are harsh especially in Finland as the temperature varies according to the outside temperature during the year, which may cause problems for the electronics equipment in the machines. The moisture variations are wide but smooth due to the roof and wall of the

storage and usually there is also a water barrier in the floor. However when the working season becomes, the machine should work reliably.

The total production volumes of agricultural implements are small in Finland. The sizes of production series are small because many different product variations are available. Therefore the control system for implements needs to be low cost. In addition to real-time requirements, working security and reliability are needed as well.

The goal of Agrix-project is to develop a prototype of an open, generic and configurable automation system platform for tractor – implement systems. The platform should be easily configurable to different implements and it should have configurable remote diagnostics and maintenance functions. Currently, the commercial control systems for agricultural machines use tailored embedded software. Their configurability varies from non-existent to very limited. The software is implemented with C or assembler programming language. This kind of approach requires large series in order to be profitable.

2.1 Tools for control software

The configuration tools should support the new standards for automation or real-time software development. One possible standard could be the IEC-61131, which has established itself as a standard notation for developing PLC-type industrial applications. A good text book about the use of IEC-61131 has been written by Lewis (1998). However, the functions of the agricultural implants can be quite complex to be implemented easily with this logic standard. The emerging IEC-61499 function block standard tries to address these limitations and seems to be sufficient software development platform for modern automation applications, see references Lewis (2001) and Christensen (2000). However, the commercial support for this automation standard is still quite limited. Constellation development tool set by RTI Ltd (2003), in which applications are defined or configured with real time UML, is used as software tool for the next version of Agrix-system. UML standard is widely used for software development in general, see e.g. OMG (2004). It has also extensions for real-time systems. With Constellation it is possible to make function blocks, which can be quite similar with IEC-61499 function blocks.

2.2 Communication standard for Agricultural Machinery

The commercially available control systems for agricultural machines have been mostly incompatible. There have been only national standards for communication between tractor and implement control systems. The lack of an open international standard is alleviated by the new ISO 11783 standard, named as "The New Standard for Agricultural Machinery", see e.g. VDMA (2001 and

2002). The standard is based on national standards, both German DIN-9684 and American SAE-J1939, so the standardization process can be said to be global and therefore believable. The standard contains now 7 final parts and at least 6 parts are still on development. The ISO 11783 standard, also known as ISOBUS, has CAN-bus at the physical layer and medium access control. The communication between electronic control units (ECUs) connected to the bus (tractor, terminal, implements, task controller, positioning device, file server) is going to be standardized and also the communication between the control system and the farm information management system.

ISO 11783 Part 12 deals with diagnostics. This work is in early phase and will contain definition of external diagnostic system connected to the ISOBUS CAN. The research in Agrix-project on remote diagnostics and maintenance could be utilized in this standardization work in later phase.

2.3 Precision farming

Precision farming also plays an important role in Agrix-project. Precision farming means that local variation in soil and other condition in the field are taken locally into account by changing certain feeding set points and working parameters to planned optimal local values according to the position measurement in real-time.

2.4 Telematics and remote maintenance

Remote fault diagnosis and maintenance in agriculture was already introduced above. It is dealt with more in detail in following sections, particularly section 5.

2.5 Positioning

The accuracy of basic GPS-receiver is suitable for precision farming purposes, but more precise positioning is needed for navigation and steering with autopilot. The precision of standard or differential GPS can be improved on the basis of dynamic vehicle model and additional measurements with cheap sensors which measure local properties, like land radar measuring velocity, inertial navigation sensors, i.e. acceleration sensors and gyroscope, electronic compass and velocity and direction of machine wheels. By combining these different measurements using model based sensor fusion, it's possible to improve the precision of positioning.

2.6 Wireless communication

Wireless communication especially between tractor and implement could reduce the connector problems, which has been found to be one of the main reasons for electronic failures in agricultural machines. Power cables are needed in any case, of course. Commercial large volume WLAN (IEEE 802.11b) modules are cheap, in some cases even cheaper than

implementing the physical layer with cables and connectors.

2.7 Driving lines and field traffic

Optimal planning of driving lines and field traffic is a little bit separate methods research area. The aim is to plan the movements in fields optimally, the criteria contain time, the usage of fuel, distance travelled, the quality of work result, soil compacting, etc. The conditions are the field (e.g. size, shape, obstacles, entrance, exit), and the machine (e.g. task, width, agility, tank size).

3. AGRIX FAST-PROTOTYPE

Agrix-project was started effectively May 2003. It was decided to make a fast-prototype of automation system for agricultural implements in order to get familiar with specific problems occurring in agricultural machines. All the requirements set in plans were not included into fast prototype requirements. One of the most important requirements in the whole project, the configurability was dismissed in fast-prototype. The demonstrated fast-prototype version was not a generic control system, easily configurable with high level tools in connection of different implements. Only one implement case was selected and the programs were made using pure C-language. The fast-proto was operational in autumn 2003 after four months of hard work.

The machine platform selected for fast-prototype was a combined seed and fertilizer driller, shown in Figure 1. The tractor was equipped with ISOBUS compatible tractor ECU.

Fig. 1. Combined seed and fertilizer driller.

The existing control system was removed and some measurements added. For example, the levels in the two containers, for seed and fertilizer; the height of the front leveling board and the coulter position (actually pressure on the field) can be measured with additional sensors. The seed and fertilizer feeding rates are controlled to planned spatial reference points on the basis of positioning with GPS. In ISOBUS architecture, each implement has own ECU as well the tractor has own ECU. The implement controller ISOBUS ECU was implemented with a commercial CAN-controller. Controller has plenty of inputs and outputs, both digital and analog. Outputs are 12V / 2A, so most of electrical actuators can be controlled directly. The controller is based on

Motorola's 32-bit 68376 microcontroller with on-chip CAN-controller.

The controller program was written in C-language and it was compiled with gcc in Linux environment. Even if the configurability for several implements was dismissed and the program was written in C, the program was designed in an object oriented manner. The functionality of the seed drill was divided into components which each handles a separate action. In spite of pure C, object oriented design was used as much as possible for later use.

Most actuators for control of the seed drill are hydraulic, powered by tractor's hydraulic system. In the original version the hydraulic actuators were coupled in a way in order to save the number of hydraulic valves needed from tractor. Even if this coupling was made quite easy to understand, it was quite difficult to handle with computer controller. The original hydraulic system in the drill was replaced with a new electro-hydraulic valve block, with which all hydraulics could be controlled separately.

The architecture of Agrix fast-prototype, shown in Figure 2, was designed according to ISOBUS, ISO 11783. Three physical ECUs were connected to CAN with ISOBUS upper protocol layers: Tractor ECU, Implement ECU, Terminal (Virtual Terminal VT + Task Controller TC). Laptop-PC was also connected to bus, for logging and analyzing purposes. At this phase, any separate ISOBUS VT (Virtual Terminal) was not used.

In industrial automation, standard, inexpensive PCs are used as HMI (Human Machine Interface). Accordingly, it was tested if an inexpensive standards PDA or handheld (HP iPAQ) could be used as HMI or user terminal for the implement. The connection to ISOBUS were made using CAN-PC Card. The programs for PDA were made with Microsoft tools for PocketPC 2002 operating system. The user interface in handheld is quite small and limited if something else is going to be done at the same time (like drive a tractor).

To improve the operability an external control keyboard was connected into the PDA. Both the VT (Virtual Terminal) and TC (Task Controller) software were running in the PDA. Afterwards it was discovered that some of the processing power problems were due to the standard, non-real-time PDA operating system. The CAN-driver also required plenty of processing power when the bus traffic was high. The CAN-driver for PocketPC did not support hardware filtering of incoming messages. The GPS receiver was connected to the PDA with Bluetooth™. For most of the time the Bluetooth worked fine. But occasionally the connection failed and re-establishing it required multiple reboots.

The standard office PDA with cover and external keyboard are in shown in Figure 3. There are also

more rugged PDAs for outdoor use, which would be more suitable for this kind of use.

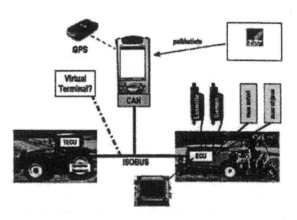

Fig. 2. The ISOBUS architecture in Agrix Fast-proto system.

Fig. 3. The HMI implemented with standard PDA and external keyboard.

The HMI software contained eight different graphical displays for monitoring and operating the machine. For different alarm and emergency situations there is a separate display. In manual mode, all actuators can be controlled separately: the control functions of the coulter unit attached to the 3-point hitch of the hopper unit, the height control of the levelling board, tramline functions, marker functions, hopper capacity, blower monitoring. The automatic mode contains several sequences: starting for starting the run, for ending the run, marker control and tramline control. As example only one display is shown here. The main operation display, illustrated in Figure 4, shows velocity, drilled area, the hopper capacity, marker and tramline state, and easy entrance to the most common functional displays. Almost all imaginable operations are implemented, and the number of measurements is the widest reasonable.

4. THE FIELD EXPERIMENTS

The Agrix fast-prototype was finally tested in real drilling of wheat in the experiment field of 6 hectares. There were some problems in localization and for this reason the feed rate was not always controlled to right values. The PDA software required occasionally more processing time than what was available. However, for the most time the system worked as planned. The experiments were executed at the research farm of the Agricultural Engineering sub-unit (VAKOLA) at the MTT Agrifood Research Finland.

5. REMOTE MAINTENENCE

In telematics, wired or mobile communication services, private or offered by communication operators, are utilized in remote operation or support of industrial process, machines or devices. Telematics applications, remote fault diagnosis, service and maintenance were at first implemented in connection of space and military technology but civil applications in connections power plants, paper machines, community technology and e.g. in elevators emerged in the 1990's as part of so called extended product concept. There exists commercial remote service centres for monitoring and diagnosing over different kind of networks not only paper machines and power plants, but also certain quite expensive agricultural machines, like e.g. potato harvester Grimme (2003) and forestry machines Ponsse (2004). In this project, remote maintenance will be realized in connection of low cost automation in agricultural implements.

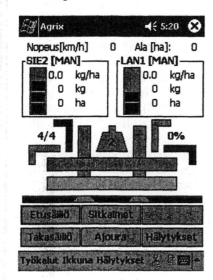

Fig. 4. The main operation display.

5.1 Communication link and service provider

In Europe, the only way to realize a communication link from the machine unit in the field to the service provider far away is via the public mobile communication networks, data transfer protocols over GSM and later UMTS; WLAN will probably never cover the whole countryside due to the quite short coverage.

An experimental test system has been realized. In this scheme, the communicating element in Agrix-system is a PDA (Compaq iPAD), which is either connected via Bluetooth with a GSM mobile phone (Ericsson R520) or alternatively has itself GSM/GPRS PCMCIA-card. The GSM device communicates over circuit based GSM-data or packet based GPRS with the service provider, a PC, via internet as client-server using sockets as interface. The data needed in remote maintenance can be transferred between PDA and service provider over GSM and Internet.

This service provider could analyse the faults on-line based on the state information sent automatically by the machine control system. After analyzing the data the service provider can advice and support the farmer in repairing the machine. In the fast prototype these remote services were not yet implemented. This will be done in summer 2004 with the basis Agrix-system realized in spring/summer 2004.

It is worth noting that remote Internet servers communicating via mobile networks with Agrix-system will be demonstrated in the context of quality control and traceability of the agricultural operations. This system can be likely utilized as the platform of remote maintenance as well. Many farmers probably don't like this kind of centralized "big brother" control of agricultural operations; most farmers will plan and record the executed operations in their own farm PC. However, dislike of centralized quality control systems does not likely mean dislike of remote maintenance, if it works well and proves to be useful. The key feature in making this kind of commercial remote maintenance profitable is that the system becomes a real commercial generic platform, which can be configured to several different implements, in order to increase the amount of installed units high enough.

5.2 Fault diagnosis and telematics in Fast Proto

The realized fast prototype contained some functions for simple fault diagnosis and fault tolerance. All measurement and actuator type were analysed in order to find certain systematic ways to utilize logical redundancy i.e. to calculate or reason the status on the basis of other sensors, just monitored elapsed time or just the situation.

The Agrix Fast Proto contained checks and alarms for (interrupt of) data transfer in CAN-based ISOBUS, (low limit of) angular velocity of the blower for pneumatic transfer of seed and fertilizer, (low limit for) rotation of feeding axels, (proper control of) feed rates of seeds and fertilizer, and discrete levels of hoppers for feed and fertilizer.

In automatic mode, the limits or discrete positions of certain linear hydraulic actuators were checked on the basis of proximity sensors. The time intervals required for movements are recorded. If certain limit sensor breaks down, the recorded times can be used to stop the motion if the feedback from the sensor has been detected to be broken down. The system is planned to be fault tolerant so that a single broken sensor does not paralyse the whole system; the fault can be circumvented somehow.

5.3 Fault diagnosis methods

Different kinds of dynamic models based fault detection and diagnosis methods have been studied throughout in the process automation context, see e.g. Patton *et al* (2000). The application of these methods in working machine context is difficult, because the monitored subsystems are somehow too simple SISO systems in order to utilize real logical redundancy on the basis of MIMO dynamical models. In the same way another main approach in fault diagnosis, based on multivariate statistics, see e.g. Chiang et al (2001), PCA PLS etc. seem to be "heavy-duty" in this application. In this drill case, quite simple scalar statistical analysis and classification to temporal feature patterns can be applied on analysis of possible faults in the analog measurements. The most important variable on which one could apply model based dynamical methods is the behaviour of oil pressure in main line measured during hydraulic valve control, which is measured and available. The behaviour of oil pressure is very context dependent but it surely contains information about certain faults. In the position control of hydraulic cylinders the hits on the limits can be easily detected on the basis of oil pressure peaks if the situation is known. These will be studied. Systematic and simple procedures for remote fault diagnosis and maintenance will be developed and tested. It is interesting to see what kind of measurement data and state information is the best for remote fault diagnosis and maintenance. The methods should be reliable and there should not be wrong alarms nor undetected faults.

Schedule based maintenance procedures should be developed for the start of the season. It is maybe quite difficult to innovate condition based method for real remote maintenance; changes in elapsed times may reveal some forthcoming faults in certain movements before they occur. Fault based methods are the last chance, but faults have already taken place.

6. CONCLUSIONS

The ISOBUS standard will be very important as open communication standard for machines in agriculture. It will solve real incompatibility problems. The ISOBUS seems to be truly widely supported and international, but standardization process in not yet completed. In all communication standards for industrial automation the tools for configuring monitoring and control functions are essential part of the toolset. The ISOBUS does not support configurability at all. The ISOBUS standard contain only communication protocols and format specifications for important variables in agriculture.

In the fast-proto system, the biggest problems occurred with the PDA. Its display is quite small for use as HMI. The cheap external keyboard was a bit too vague for this purpose. The processing power of the PDA was quite limited for this kind of use and real-time problems emerged in some situations. The CAN-card needs also plenty of processing power in order to handle the whole of bus traffic.

In remote fault diagnosis and maintenance sector, a mobile communication test system was realized. In fast proto phase the focus was on stand alone fault detection and diagnosis and fault tolerance. Remote fault diagnosis and maintenance has been planned, but it will be realized later in the next version basis Agrix-control system in summer / autumn 2004.

Acknowledgements
This work has been supported by Tekes National Technology Agency of Finland and companies Valtra Oy, Nokka-Tume Oy, Junkkari Oy, Mitron Oy, Bitcomp Oy, Kemira GrowHow Oy and Vieskan Metalli Oy. The automation group at HUT thanks for the fruitful collaboration MTT Agrifood Research Finland Agricultural Engineering sub-unit (VAKOLA), TTS Institute (Work Efficiency Institute) and Deparment of Agricultural Engineering and Household Technology at Helsinki University.

REFERENCES

ISO International Standards Organization. (1984). Information Processing Systems - Open System Interconnection (OSI) - Basic Reference Model, International Standard, ISO 7498 / CCITT X.200, October 1984, Geneva http://www.iso.org/

ISO (2003). ISO/WD 11783-1. Tractors, machinery for agriculture and forestry - serial control and communication network Part 1: General Standard. Revision: 2002 July 10. ISO/ TC23/ SC19/ WG1/ No. 277/02E.

Lewis R.W. (1998). Programming industrial control systems using IEC 61131-3, IEE Control Engineering Series 50

Lewis R. (2001) Modeling Distributed Control Systems using IEC 61499, IEE Control Engineering Series 59,

Christensen J (2000). Basic Concepts of IEC 61499, 24.10.2000, http://www.holobloc.com

Object Management Group (OMG). Unified Modeling Language UML. http://www.uml.org

VDMA (2001). *ISOBUS Communication System - The New Standard for Agriculture*. Messuesite. VDMA, Landtechnik, Frankfurt am Main, Saksa. 4+8 s.

VDMA (2002). *ISOBUS Spesifikation, Implementation Level 1*. 2002-04-25. VDMA Landtechnik, Frankfurt am Main, Saksa. 18 s.

RTI Ltd (2003): Constellation UML Guide. http://www.rti.com/

Grimme (2003): http://www.grimme-online.com

Ponsse (2004): http://www.ponsse.fi

Patton R.J., Frank P.M., Clark N.R. (eds)(2000). *Issues of Fault Diagnosis for Dynamic Systems*. Springer, 570 pages.

Chiang L.H., Russell E.L., Braatz R.D. (2001): *Fault Detection and Diagnosis in Industrial Systems*. Springer, 275 pages.

Copyright © IFAC Telematics Applications in Automation and Robotics, Espoo, Finland, 2004

ELSEVIER
IFAC
PUBLICATIONS
www.elsevier.com/locate/ifac

REMOTE CONTROLLABLE VIBRATION DAMPING PLATFORM

Otto J. Roesch, Alexander Prusak, Hubert Roth

Institute of Automatic Control Engineering (RST)
University of Siegen, Germany
hubert.roth@uni-siegen.de

Abstract: In the study of engineering education, presently a change of learning and teaching happens by the use of new media. The reason for this process is the implementation of tele-laboratories into the curricula of automatic control engineering and mechatronics education. Real existing laboratories can get controlled remotely over the Internet with a web browser by students. An accompanying usage of these lab experiments during the lectures as well as an independent use is considered in the theoretical teaching units and tutorial. The article discusses the development, access and finally the usage of such online experiments. *Copyright © 2004 IFAC*

Keywords: Remote Control, Telematics, Teleoperation, Telerobotics, Telecontrol, Learning Systems, Control Education, Laboratory Education, Automatic Control, Flexible Arms

1. INTRODUCTION

Currently a change in the engineering education with the usage of eLearning systems takes place. The accompanying education of conventional lectures together with remote controllable experiments strengthens the practical oriented and industry-relevant education at our Universities. Studied material can directly be applied and tested on real experiments, controllable over the internet. The experiments are physically distributed in different Universities, all online accessible over an internet platform. Students are not dependent any more in location and time constraints, because the experiments can get accessed at any time and independently without any tutor directly over a web browser.

The available eLearning units can be divided into different access methods, where offline accessible teaching units are called Computer Based Training (CBT) and consist of interactive self-learning programs, the just online available eLearning material is called Web Based Training (WBT) and consist e.g. of self-learning programs with tutorial accompaniment, chats, forums, FAQs and in our case with remote controllable real existing online experiments (Langmann and Hengsbach, 2003).

The University of Siegen is involved in different national and international projects, which deal with the development of remote laboratory experimentation via Internet. These experiments are all based in the field of automatic control engineering and mechatronics. The national participating projects are LearNet (learning and experimenting on real technical plants, www.learnet.de) and Learn2Control (project oriented multimedia learn environment for control engineering, www.learn2control.de), the international projects are IECAT (Innovative Educational Concept for Autonomous and Teleoperated Systems, http://www.ars.fh-weingarten.de/iecat) and TEAM (Tele-Education in Aerospace and Mechatronics Using a Virtual International Laboratory, http://www.ars.fh-weingarten.de/team). Developed pedagogical concepts and course units offer innovative laboratory experiments with hardware controllable via internet. By including a wide audience of students at the partner universities, the exchange of student and faculty members is guaranteed. Thus students profit from this transatlantic tele-education approach by training in advanced technology in an international environment.

The experiment explained here is a test-platform for vibration damping and high-precision positioning of a flexible structure. The aim of this online experiment is to study the identification and control of the mechanical vibration in flexible structures. The students have to identify the system dynamics and test different control strategies. Finally, they have to evaluate the quality of the control algorithms. In order to provide a handy and convenient usable laboratory experiment, a complete educational unit has to be provided. To help students to get into the appropriate field, links to general tutorials as well as a set of documentation to the particular experiment is accessible over the internet. Before the student can execute the experiment, a knowledge check has to be passed. This consists of online questions like multiple-choice.

The access to the remote experiment is done by an eLearning web-portal. This includes a booking system, in which the registered students can reserve time slices for the online laboratory experiments.

2. CONTROL OF FLEXIBLE STRUCTURES

In the design of spacecraft, aircraft, and even building, the flexibility of structural elements is of concern. This is especially pronounced in space structures and aircrafts where large size coupled with lightweight materials emphasize structural flexibility. The control of such structures becomes problematic due to this behaviour. For rigid structures accurate theoretical models can be obtained. However, for flexible structures modelling, simulation, and controller design is difficult at best. To familiarize students with these control techniques, an experiment dealing with the control of flexible structures has been developed at the Control Engineering Department in the University of Siegen. The focus of this experiment is an aluminium rod, suspended on a motor. Its sensors and actuators can be controlled via the Internet. The rod's length coupled with its small cross section makes the system quite flexible. The purpose of this lab is to provide a test platform for students to analyze structural vibrations, model system behaviour and design controllers. They can implement their work on an actual physical system via the Internet or locally in the laboratory.

2.1 Test Facility Hardware

The main feature of the "Swinging Rod" test facility is shown in Fig. 1. It uses a 1.6 meter long aluminium rod with a small cross section (4x10 mm) as the flexible component (Muenst and Roth, 2001). An excitement of the rod is possible in one orthogonal direction at the upper end of the rod. The system is equipped with a DC-Motor as an actuator, used for disturbing and afterwards to control the rods movement. The following two sensors are used for the vibration measurement (Irwin, et al., 2001):

A PSD (Position Sensitive Detector) sensor is used for the position measurement at the tip of the rod. It is an opto-electronic device which converts an incident light spot into continuous position data.

Advantage of this sensor is the high resolution, a fast response and a very good linearity for a wide range of light intensities. In principal the PSD is a photo diode with an illuminating sensitive surface. On the chip, the p-surface of the is light-sensitive, whereas the n-surface is continuous metallised, only on the edges of the p-surface are metallic electrodes, (Fig. 2). If a point of light meets the PSD charge-carrier, it causes a current flow to all electrodes on the edges. The unlighted areas at the chip act like resistors, therefore the ratio of the currents on the electrodes depends on the position of the incident light spot. The current-ratio is independent of the incident light, whereby a stable measurement is done. A lens is fixed at top of this photo–element; by this an extension of the working range of 20 centimetres in diameter can be achieved. The PSD is located on the floor, perpendicular to the rod. An infrared diode, fixed at the tip of the rod, beams down towards the sensor.

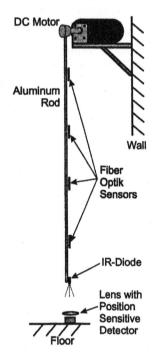

Fig. 1. Schematic Drawing of the "Swinging Rod"

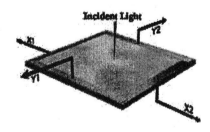

Fig. 2. Design of the Position Sensitive Detector

The second sensor measures the angular position of the motor shaft, therefore a built in 4–quadrant encoder inside the motor is used. An analogue amplifier drives the motor and delivers the necessary current. In comparison to PWM (Puls Width Modulation) amplifiers, no high frequency noise is inside the system and disturbs the very small analogue sensor signal from the PSD.

2.2 Software Structure

The software is divided into two main parts, one is the software to communicate directly with the hardware over a data acquisition board, and the other program establishes a remote control over the internet (Fig. 3).

The control software to communicate with the hardware is divided into four sub-modules.

- The Java Server handles on one side the communication between the server and the client (student) over the internet (remote control software), and on the other side to get access towards the hardware with a DLL library.
- The DLL library is an interface between the Java server application and the Matlab program, to enable the communication between these applications.
- Matlab/Simulink together with the Real-Time Workshop is first used during the development phase to create the real-time code. During the execution of the experiment, Matlab starts and stops the real-time process and saves the sampled data over the WinCon application.
- The WinCon software controls the data acquisition board (both from the Quanser company) over the RTX (real time kernel) from VentureCom.

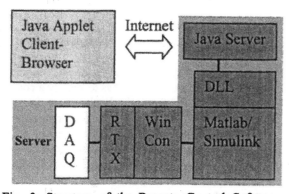

Fig. 3. Structure of the Remote Control Software (DAQ – Data Acquisition Board)

The remote control software is implemented in Java, whereas on the server-side a Java application and on the client (student) side a Java-Applet is developed. The Java-Applet is integrated into a web page and is accessible inside a web browser. No special software is needed on the client side, except a web browser which allows Java-Applets. The communication between the client and the server over the internet is managed by a socket communication. The control signals form the client to the server and the feedback information backwards get transferred over this connection (Schilling, 2001). The sensor data as displayed in Fig. 4 are transferred to the student after the particular real-time process is finished. Due to the fact that this is a fast vibrating system, the sensor signals get transferred after the realtime process is finished.

The movement of the rod is also observed by a web camera to get a better expression of the oscillating rod and to see how well the controller works.

2.3 The Accomplishment of the Swinging Rod Experiment

The students get access to theoretical teaching units in order to process the tasks. These units are available in HTML over the internet and do not have any restrictions in the access time. The content of the educational unit is to study the identification and control of the mechanical vibration in flexible structures.

First the students have to get familiar with the mathematical description of flexible systems, especially with the Lagrange approach. Whereas they first have to find the energy equations of a flexible beam, then complete the Lagrangian equation (James, et al., 1993; Meirovitch, 1986; Feliu, 1997) and derive the complete equation set of movement of the system. Afterwards the motor-equations get implemented and the differential equations must be transferred into a forth order state space form (Gawronski, 1998).

To complete the model, the natural frequency of the swinging rod must be measured. This is done with the impulse response of the link, whereas the oscillating frequency represents the natural frequency of the system. A torque impulse is applied to the motor, so that the rod is decaying with its natural frequency. The experiment must be accessed with the remote control in this step. More advanced students in control engineering get an additional feature, whereas the sampled data can be downloaded and system identification with the sensor data can be performed. For this step, software like the System Identification Toolbox under Matlab would be necessary. This subtask is not compulsory for successfully performing this lab experiment.

In the next step, the student has to enter the derived model as linear state space description into the remote control interface to verify it. The system matrix A and the input matrix B must be inserted. An overlapped plot of the real data with the simulated data, as an impulse response, is shown in the web browser to the student (Fig. 4).

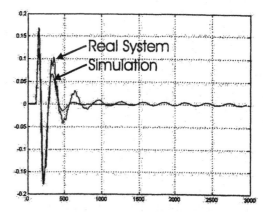

Fig. 4. Shown Diagram of the Remote Control Interface Inside a Web Browser

Aim of the active vibration damping is to set the rod as fast as possible in quiescent after an oscillation; especially the tip of the rod must be calm (Rahn, 2001). The control is implemented with the LQR control approach, whereas an introduction of how to design the weighting matrices for the LQR control, based on the Riccati equation, is explained in the teaching units. The students have to design the weighting values in such a way, that the decaying oscillating time, when the controller is switched on, goes to zero. The weighting values get inscribed directly into text fields inside the remote control window. These values get transferred directly to the server, whereas the LQR command under Matlab (on the server side) gets executed for receiving the control vector. For this calculation, the model derived by the student from the previous step is used. For the control procedure, first the rod gets excited with a sinusoidal signal for around 6 seconds, till it is nicely oscillating. Then directly following, the disturbance signal is switched off and control is activated. The decaying time shows now, how well the controller works. Afterwards a plot of the sampled sensor data of the motor angle and the tip position is created and loaded automatically into the right side of the browser window in figure 4.

A comparison between the real controlled system and a pure simulation with the students' model verifies the mathematical model with its parameter identification further more. All inserted data with its created plots get saved on the server together with the student identification number for an evaluation.

3. THE eLEARNING PORTAL

The eLearning portal intends, that several providers can offer their own teaching modules implemented on one or several own web servers. The hosting and maintenance of teaching modules lies within the responsibility of each provider. The user registration and administration belongs to the supervisory of the eLearning portal. Such a distributed system of web resources requires high demand regarding security and access rights on the portal and the external modules.

The eLearning portal was built with the free portal software PHPNuke (PHPNuke). The eLearning portal integrates several teaching modules and additionally disposes portal service modules (news, forums, and glossary), user management, teaching module management, booking system and security system (Fig. 5).

The contents of the portal are given in three different types of teaching modules:
- The overview and introduction give general information for all unregistered users about the learning unit and the online experiment.
- The learning unit will provide automatic control theories to the student as well as prepare the user

for the online experiment. It includes scripts, manuals, literature, simulations, examinations and multiple choice tests.
- The online experiment is a real hardware system controllable via internet. Learning units and online experiments can only be used by registered users.

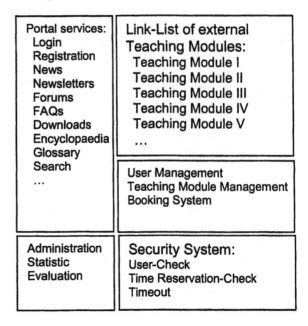

Fig. 5: Overview of Functions and Components of the eLearning-Portal

3.1 The Booking System for Time Reservation

Standard users only get access to general information. For user registration each student has to specify a coach of his own university and the desired online experiment. These registration data will be saved in the portal-database and in addition they are sent to the coach by email, who has to confirm the user registration. Now the user has full access to the learning units, can book time slices for laboratory use, (see Fig. 6) and gets a direct link to the online experiment.

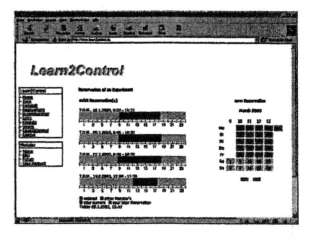

Fig. 6: Listing of all Existing Time Reservations of one Online Experiment (middle) with the Possibility of new Bookings (right)

3.2 The Portal Security System

The eLearning web pages as well as the provider web pages are accessible after authentication inside the portal. But these web pages could also be loaded by bookmarks or links from other pages. By using this mechanism each web user could get access to the online experiments.

While inside the eLearning portal, the authorized user data will be stored in an anonymous way, i.e. user- and session-identification-number. This information is needed for further work with the portal web pages and also the providers' web pages. Before the user switches to a restricted provider's web page, the last portal web page includes a JavaScript, the student-ID gets checked with the booking system, if the access can be granted, see Fig. 7. After correct identification, the student gets the access to the appropriate teaching units.

Now the user can switch to the provider's web pages. These pages are presented inside a frameset. These frames include a portal user check Java-Applet and JavaScript. These resources handle the security system as user and reservation check, timeout, etc.:

- The provider differs between web pages of learning units and web pages of experiments. These pages will be managed differently by the portal. The type of the web page is set by a JavaScript variable.
- In regular time intervals the user-check-applet verifies the user-id and session-id from the portal and compares it.
- Experiment web pages will be checked for the reserved time slice.

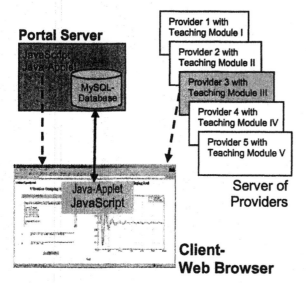

Fig. 7: Security System of the Client-Provider-Portal-System with User Check, Time Reservation Check and Timeout

⬅ - ▶ Webpages together with the Java-Applet and the JavaScript are send to the Client

⬅⬅⬅➡ Communication between the eLearning-Portal, and the client's browser

4. EXPERIENCES

By the present implementation around 50 students tested this platform as a laboratory experiment from outside. At the end of the online execution, a multiple choice questionnaire and hand written tasks like the model derivation and controller design were delivered. Due to the fact that this is not a simulation, the students are more serious and really try to design a good controller, but also a lot tried to destroy the system by using badly designed control parameters.

A statistical analysis is in work and will be given in the final paper, to publish for this IFAC conference.

5. CONCLUSION

The developed work of the vibration damping experiment "The Swinging Rod" enables the access by a lot of students from different universities. By sharing physically real remote controllable experiments by many universities, a huge financial benefit is achieved. During the development process, different hard- and software solutions were tested. Due to timing issues and an unstable behaviour of the real-time process, the present system was chosen.

The eLearning portal handles the teaching units, whereas a teaching module can be separated into three parts: general information, learning unit and experiment. General information and learning unit can be offered as pure internet media. But experiments represent real laboratory equipment, which can be driven remotely via Internet. In particular the external hosting of teaching modules needs special demands for the eLearning portal. An extensive booking system provides a secure administration of users and resources.

One main platform for accessing such online experiments offers physically real eLearning experiments to a wide audience. All experiments inside this portal are developed in English can be accessed from anywhere in the world.

6. ACKNOWLEDGEMENT

The support of these projects by the BMBF (German Federal Ministry of Education and Research), UVM (Universitätsverbund MultiMedia des Landes Nordrhein-Westfalen) and the European Commission is gratefully thanked.

7. REFERENCES

Feliu Batlle, J. J.: Desarrollo de Modelos Dinámicos Computacionalmente Eficientes para Robots Flexibles de un Grado de Libertad. Aplicación al

Control Adaptativo. Doctor Thesis, 1997. Escuela Técnica Superior de Ingenieros Industriales, Universidad Nacional de Educación a Distancia, Madrid.

Gawronski, W. K.: Dynamics and Control of Structures; A Modal Approach. Springer Verlag 1998, p.11 – 27.

Irwin R. D., Adami T. M., Roth H., Münst G., Roesch O.; Sensor and Control Concepts for the Internet-Based Flexlab Experiment; Workshop on Internet Based Control Education, IBCE'01, December 2001, Madrid - Spain, page 231 – 234

James, M. L.; Smith, G. M.; Wolford, J.C.; Whaley, P. W.: Vibration of Mechanical and Structural Systems. Harper Collins College Publishers 1993.

Langmann, R., Hengsbach, K., (2003). E-Learning & Doing automation. ATP, 45, page 58 – 66.

LearNet, URL: www.learnet.de

Meirovitch, L.; Elements of Vibration Analysis, McGraw-Hill Book Company, ISBN 0-07-041340-1, 1986

Muenst, G., Roth, H., Sensoring and Control for Flexible Structures. Workshop on Tele-Education in Mechatronics Based on Virtual Laboratories. Weingarten, Germany, 2001. ISBN 3-9255359-003, p. 60.

PHPNuke, URL: http://www.phpnuke.de

Rahn, Chr. D.; Mechatronic Control of Distributed Noise and Vibration, Springer, 2001

Schilling, K.: Remote Sensor Data Acquisition and Control. Smart Systems and Devices, Tunisia, 2001

Copyright © IFAC Telematics Applications in Automation
and Robotics, Espoo, Finland, 2004

www.elsevier.com/locate/ifac

REMOTE EXPERIMENTS IN CONTROL EDUCATION

Mikuláš Huba, Pavol Bisták and Katarína Žáková

Dept. of Automation and Control
Faculty of Electrical Engineering and Information Technology
Slovak University of Technology
Ilkovičova 3, Bratislava, Slovakia
huba@elf.stuba.sk

Abstract: This paper reports about experience in building Control Laboratory
with remote access via Internet and in organizing the learning process with
the aim to foster an active and creative learning by doing. *Copyright © 2004
IFAC*

Keywords: Control education, Multimedia, Laboratory education, Resource
Based Learning, Tele-experiments, Flexible Learning

1. INTRODUCTION

Two facts are dominating the control area, research
& education today: First, the quantity of information
available in the world is exponentially growing and
also the rate of this growth is accelerating. The new
knowledge-based global society is one, in which
(Resta et al., 2002):

• the world's knowledge base doubles every 2–3
years;
• 7,000 scientific and technical articles are published
each day;

Each researcher and teacher is then faced with a
strategic question - how to remain up-to-date in so
rapidly changing world?

The second modern advance is the new capacity to
communicate among people of the world. In the
control area, we can use this new capacity to
communicate the experimental process data, too.

Improving the quality of research & education
through the diversification of contents and methods
and promoting experimentation, innovation, the
diffusion and sharing of information and best
practices as well as policy dialogue are
internationally recognized as strategic objectives. In
this paper, we are going to present several results
achieved in building up Internet based tele-
laboratory. In the research area, the tele-laboratory
can be primarily used for promoting experimentation,

innovation, the diffusion and sharing of information
and best practices among networked partners and
communities.

Educational systems around the world are under
increasing pressure to use the new information and
communication technologies (ICTs) to teach students
the knowledge and skills they need in the 21st
century. Several documents, as e.g. the 1998
UNESCO World Education Report "Teachers and
Teaching in a Changing World" describe the radical
implications ICTs have for conventional teaching and
learning. They predict the transformation of the
teaching-learning process and the way teachers and
learners gain access to knowledge and information.

With the emerging new technologies, the teaching
profession is evolving from an emphasis on teacher-
centred, lecture-based instruction to student-centred,
interactive learning environments. Tele-laboratories
have the potential to transform teaching and learning
processes and to transform the present isolated,
teacher-centred and text-bound classrooms into rich,
student-focused, interactive knowledge
environments. Environments in which students are
engaged learners, able to take greater responsibility
for their own learning and constructing their own
knowledge.

The learning environment (Resta et al., 2002) that
may be derived from this view of the learning
process is shown in Fig.1.

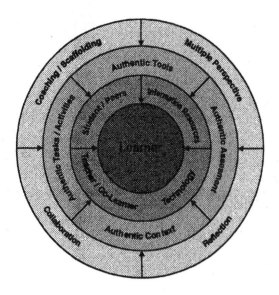

Fig. 1 Student-Centred Learning Environment

The student-centred environment illustrated in this figure shows that the learner interacts with other students, teachers, information resources, and technology. The learner engages in authentic tasks in authentic contexts using authentic tools and is assessed through authentic performance. The environment provides the learner with coaching and scaffolding in developing knowledge and skills. It provides a rich collaborative environment enabling the learner to consider diverse and multiple perspectives to address issues and solve problems. It also provides opportunities for the student to reflect on his or her learning.

2. VIRTUAL LABORATORIES

The rapid development in the information and communication technologies enables to increase the diversity of means used in teaching and education. Internet technology offers more ways how to reach students, transfer knowledge and develop skills. For the branches of education that require the development of hands-on abilities with laboratory equipment – namely in engineering – the possibility to tele-operate equipment has opened new horizons for tele-education. Virtual laboratories aim at providing via the Internet the hardware components and associated practical skills that complement WEB-based course material (Schmid, 1999).

The necessity of learner's experimental work in laboratories represents one important aspect of engineering education. Students solve practical problems and gain experience and practice needed for their future career. Introduction of real systems increases transparency of the solved examples.

The main motivation for using of plants in the educational process is clear physical "visibility" of the controlled dynamics, and also the necessity to

exercise all design steps starting with the plant identification and ending with the evaluation of the control results achieved with the particular model. In fact, the model by itself can be presented as a virtual device whose dynamics is described by differential equations or as a real plant. With the development of new computer technologies, an interactive multimedia programming language JAVA, and the WorldWideWeb, it is now possible to simulate a virtual model on a computer and to offer it as an animation to students in frame of "virtual laboratory" via the WWW or CD-ROM. However, use of the animation models cannot fully substitute the work with real physical plants that always demonstrate some unmodelled dynamics, parasitic noise, parameters fluctuations, etc. Unfortunately, the number of students is often high in comparison with the number of available real plants. Building remote laboratories that gives learners access via Internet represents a possible solution of this problem (Fig.2).

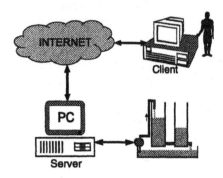

Fig. 2. Experiment control with the real plant through Internet

Expensive and elaborate experiments can be set up once and can be made available all over the country or even the whole world via Internet. Experiment-oriented problems can be offered without the overhead incurred when maintaining a traditional laboratory.

In preparing a tele-experiment, it is important to take into account that students should be actively involved in. They must be able to change parameters and to collect measurement data. As far as possible, the processes should be observable through the Internet. So, also the time for conducting an experiment must be short enough to be made in a session of 5-15 min and to avoid longer waiting queues.

One of the main streams in the development of the contemporary control theory is obviously represented by the nonlinear control. In order to be able to "touch" and evaluate the newest control techniques, several plant models with typically nonlinear behaviour have been built.

The aim was to offer to students various types of systems regarding their physical nature and their dynamics. In following it is possible to find short description of the available systems.

3. CONCEPT OF THE SYSTEM

The concept is based on the existence of classical control engineering laboratories with experiments carried out in the face-to-face form. The already built infrastructure is going to be used in a more open environment. In the control education Matlab software is usually used for implementing various control strategies. Due to different toolboxes (like Real Time Workshop, Real Time Windows Target and xPC Target) Matlab with Simulink is no more only a simulation tool. Nowadays, it also enables to control real plants. There exist many experiments using this platform. Our aim was to open laboratory equipment to larger groups of students and researchers, especially those, who cannot be regularly physically present. This extension requires only small changes in laboratory hardware equipment and a little bit more programming work in software equipment (Müller & Waller, 1999).

To allow remote users to run experiments the widely accessible Internet connection was chosen. The data flow between the real plant placed in the laboratory and the remote user is handled by a client-server application that communicates through TCP/IP protocol. This is a Web-based application so the remote user can reach experiments within the familiar environment of his/her Web browser.

On the server side the application has to communicate with the Matlab that directly controls the real plant. It is advantageous, when the server part of the application resides on the same PC that controls the real plant. Then, within one operating system, the data are exchanged more easily.

The transfer channel is complicated and moreover the Internet connection is not a very reliable one. Therefore, it is advisable to divide the whole program application for tele-experiments into following parts

- Local control (verified control strategies implemented locally with the Matlab application)
- Server application (this handles data flow between the remote client and the real plant, it transfers user's commands and transmits responses)
- Client application (remote user interface that enables to control a specific real plant usually in the form of JAVA applet running within Web browser)
- Administration (an application that covers administrative issues in the case of many users for several real plants)

Next, two of these topics: the local control and the client applications will be treated more deeply.

Real plants and computers equipped with a data acquisition cards establish the prevailing structure of a Real Plant Control laboratory for the Control Theory education. A simple PC with Matlab offers a comfortable way for testing different control strategies. The results can either be simulated, or measured from the real plant. Matlab provides support for several A/D and D/A cards. For other cards it is necessary to write drivers.

In our Real Plant Control Laboratory (see e.g. Žáková et al., 2000), one can find following physical models controlled by Matlab with the AD512 and MF604 data acquisition cards: two tank system, magnetic levitation, inverted pendulum/gantry crane, helicopter rack, thermal plant, ball and beam with propeller control and mining lift.

All models can also be controlled via Internet. For this purpose some of the plants have to be modified. For instance the two-tank system was equipped with an additional set of valves that enable its reconfiguration from the two-tank system to one-tank system and vice-versa. These valves also enable simplified identification of particular system parameters. All systems equipped by incremental sensors: the helicopter rack, ball and beam, mining lift and inverted pendulum/gantry crane have to be equipped with the possibility to set remotely a reference position.

In fact, each real plant offers possibilities to demonstrate a wide spectrum of nonlinear phenomena. Students (and also researchers) can exercise complex design tasks ranging from the system identification up to the controller design and to the evaluation of the resulting control quality. Up to now, various linear controllers based on linearization around fixed operating point and nonlinear controllers (generalised exact linearization, constrained control, anti-windup PID control, or fuzzy & neural control) were implemented. Introducing tele-experiments these results can be compared with those achieved by other remote users.

Fig.3. The two-tank system

4. TWO TANK SYSTEM

The two-tank system (Fig.3) belongs to the basic plants used in control education. Its dynamics is nonlinear, can be chosen within well-timed limits, it is visually clear and enables to verify a wide spectrum of control algorithms.

In general, TCP/IP client is an application that connects to a specific port on a TCP-IP server and exchanges data either as a stream or text. It is usually realized as Java applet that enables to modify controller or simulation parameters and visualize simulation results. After being installed on a web server it can be approached by a student via Internet.

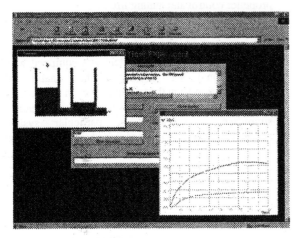

Fig.4: Java client for the two-tank system

Using this technique student can access the two-tank system via Internet and carry out remote experiments. An example of the developed Java applet with the interface, the animation window and the graph can be seen in Fig.4. It enables to connect to the real plant and to choose predefined control algorithm or to create their own one and to upload it to the server. Then to set the simulation time and to start the experiment. At the bottom of the interface window, there is one field (command line) that serves for sending Matlab commands. Results of them are displayed in the upper output window. Appearance of the graph or animation and receipt of the measured data can be checked by appropriate boxes.

5. MAGNETIC LEVITATION SYSTEM

It is an unstable nonlinear dynamic system with one input (current flow which influences magnetic field intensity) and one output (position of the steel ball, Fig.5)

The motion equation is based on the balance of all forces, i.e. gravity force F_g, electromagnetic force F_m and the acceleration force F_a

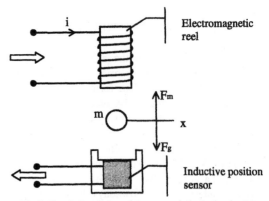

Fig.5: Sketch description of the magnetic levitation model.

The Fig. 6 shows all important windows of the Java client interface. First, it is necessary to connect to the real plant. The remote experiment is password protected and requires authentication. The first window in the Fig. 6 enables to select the control algorithm, set parameters of the experiment and run it. In the visualization box the remote user can see the animation of the magnetic levitation system. The second window displays the graph of the system output that is the position of the ball.

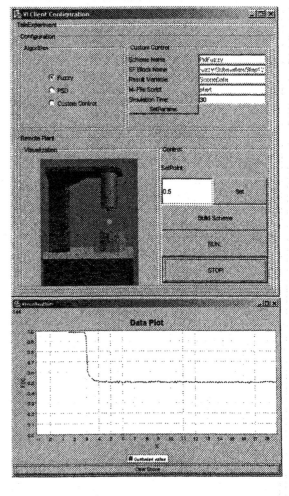

Fig.6: Java client for the magnetic levitation system

6. HELICOPTER MODEL

The helicopter rack model is representing one of the popular non-linear educational control problems (Fig7). It consists of a body situated on a base support. The body (carrying two DC motors with propellers) has two degrees of freedom. The axes of the body rotation, as well as those of the motors are perpendicular each other. The rotating propellers, which are driven by DC motors, influence both body position angles, i.e. the azimuth angle in the horizontal and the elevation angle in vertical plane. Incremental rotary sensors (IRC) measure both angles of the helicopter. The range of body rotation is ±48 degrees in elevation and ±175 degrees in azimuth.

Rotations of propellers are measured by incremental sensors, which provide frequency signal about their angular speed.

DC motors fed by power amplifiers, which are integrated on the interface board, drive the propellers. Power amplifiers are activated by output analogue signals from the superior level (PC with proper acquisition card). Main motor (elevation motion) controlled by signal 0..+10V rotates only in one direction. Side motor (azimuth motion) rotates bi-directional and is controlled by signal ±10 V. Interface board includes also current sensors of both motors. Measured currents are represented in analogue form (±5V DC).

The interface board requires supplying voltages ±12V DC, +5V DC (available directly from acquisition card of PC), and 24-30V DC for motor supplying.

The approximating helicopter model can be described as the 6th order 2x2 system.

Rotor R

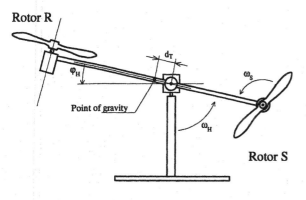

Fig. 7. Front view

Similarly to above-mentioned remote interfaces there exists a Java client for the helicopter model (Fig. 8). Remote users have to connect to the real plant, then choose the controller and set parameters (e.g. desired

values of both angles). After they start experiment they can see animation of the helicopter in the vertical and horizontal plane and at the same time the values of both angles are displayed in the form of graphs. After the end of simulation students can receive numerical data for further processing using the button Data Export. Setting of the system to the initial reference position is solved by special algorithm that runs before each experiment.

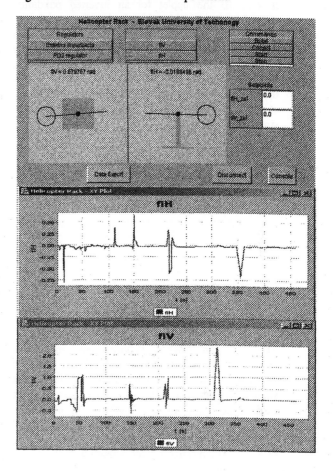

Fig.8: Java client for the helicopter system

7. INVERTED PENDULUM SYSTEM

The control of an inverted pendulum (Fig.9) is very commonly used problem. The main advantage consists in a not very difficult model of the nonlinear system. On this system it is possible to show the difficulties in controlling a plant having unstable open-loop poles.

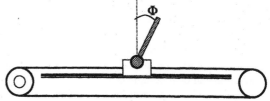

Fig.9: Sketch model of the inverted pendulum system.

The system consists of a cart, which can be moved along a metal guiding rail. An aluminium rod with a cylindrical weight is fixed to the cart by an shaft. A transmission belt to a drive wheel connects the cart. The wheel is driven by a current controlled direct current motor, which delivers a torque proportional to the acting control voltage such that the cart is accelerated.

In Fig. 10 the applet for an inverted pendulum control is shown. After its opening in an user web browser, TCP/IP client connects to the proxy server and to the S-function in the Simulink model. A user is informed about the time of connection and the connection status through short information note in the first part of the applet window ("client connected"). In this moment, everything for real-time simulation is prepared.

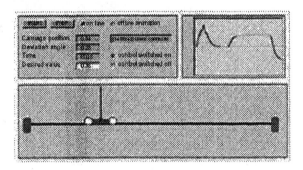

Fig. 10. Java client for inverted pendulum experiment

The user can enter a required position of the pendulum. The buttons "Start" and "Stop" serve for the control of simulation running. During the simulation user can follow numerical values of the carriage position, deviation angle of the pendulum, time of simulation and the graphical dependence of the carriage position. The simulation results can also be visualized through model animation that can run in two modes: on-line and off-line mode.

In the on-line mode the animation starts immediately after the beginning of simulation and in each moment user can see actual state of the pendulum in the animated model. The time between the real time experiment and the animation is synchronized. However, the velocity of this animation depends on the velocity of data transfer via Internet. If the transfer is not sufficiently fast, the animation is not capable to visualize the samples from the experiment on line because of big delays. Then, the off-line animation can be used. In that case the real time simulation is realized only numerically without the contemporaneous animation. After the simulation ends, measured data are sent to the client and the animation can be accomplished.

So, it means the Java client has to enable to user

- to control the process and

- to collect and visualize the data from the process

Connection between Matlab and Java client can be realized directly using TCP/IP communication.

8. CONCLUSIONS

In this paper, we have introduced basic structures and processes of the tele-laboratory. Course materials and case studies giving a guide to control particular processes (bench-mark examples) will complement them in a short future. All this is developed with the aim to enable establishment of a community approach to control problems. Students and their teaching staff give the core of this community. However, after achieving a higher degree of reliability, we intend to expand this community also by our previous students and experts from industry. We will also welcome networking with other similar laboratories and interested colleagues abroad.

REFERENCES

Resta,P. et al. (2002) *Communication Technologies In Teacher Education: A Planning Guide.* Division of Higher Education, UNESCO (available at <http://unesdoc.unesco.org/images/0012/001295/129533e.pdf>, 23.12.03)

Isidori A. (1995). *Nonlinear Control Systems*, 3rd Ed. New York: Springer Verlag.

Schmid Chr. (1999). Virtual Laboratory for Engineering Education, *ICDE conf. Vienna*, Austria.

Zimmer, G. (1995). Ein nichtlinearer Regler für ein Hubschrauber-Rack auf der Basis diffgeometrischer Methoden. *Automatisierungstechnik* **43**, 347-356.

JMatLink Version 1.00; http://www.held-mueller.de/JMatLink

Müller S., Waller H. (1999). Efficient Integration Of Real-Time Hardware and Web Based Services into MATLAB, *11th European Simulation Symposium*, October, Erlangen, Germany.

Zakova, K., Huba, M., Zemanek, V., Kabat, M. (2000). Experiments in Control Education, *IFAC ACE*, Gold Coast, Australia.

Bisták P., Žáková K. (2003). Organising Tele-experiments for Control Education. *11th Mediterranean Conference Control and Automation.* Rhodes, Greece.

Copyright © IFAC Telematics Applications in Automation and Robotics, Espoo, Finland, 2004

ELSEVIER
IFAC
PUBLICATIONS
www.elsevier.com/locate/ifac

A VIRTUAL LABORATORY FOR AN INVERTED PENDULUM AND CRANE CONTROL

I. Masár, C. Röhrig, A. Bischoff, M. Gerke, H. Hoyer

Department of Electrical Engineering
Control Systems Engineering Group
FernUniversität in Hagen
Universitätsstr. 27, 580 97 Hagen, Germany
Ivan.Masar@FernUni-Hagen.de

Abstract: A virtual laboratory allows students to gather experiences with practical on-line experiments at remote sites. Moreover, it becomes possible to share expensive equipment among several universities or education centres. Several German universities were engaged last three years in the 'LearnNet' project, in order to develop a network of remotely accessible virtual laboratories and server with various courses. One of our contributions to this project is a laboratory experiment for inverted pendulum and gantry crane control. Several new components for real-time control and Internet access have been developed and implemented. *Copyright © 2004 IFAC*

Key words: Virtual laboratory, inverted pendulum/gantry crane, xPC Target

1. INTRODUCTION

In the past few years 'Virtual labs' have been meaningful and very popular for tutorials in control theory, since they allow performing of interesting experiments dedicated to real-time control of technical systems all over the world. At our research group, several such laboratories were established, including a lab experiment for remote control of a mobile robot with omnidirectional wheels (Röhrig and Jochheim, 2000; Bischoff and Röhrig, 2001) and for Internet-based testing and programming of industrial controllers (Simatic S7 family, Siemens). At the present time, we developed a new virtual laboratory for real-time control of an inverted pendulum and gantry crane system. This lab is one of our contributions to the 'LearNet' project, in which several German universities collaborate in order to set up a common platform for on-line learning and experimentation in control theory.

2. VIRTUAL LABORATORY OVERVIEW

As mentioned above, our new virtual lab allows a teleoperated control of the inverted pendulum and gantry crane system, respectively. This system was chosen because it represents one of the most commonly used non-linear systems in control theory at the undergraduate level. The ability of this system to provide interesting and challenging experiments makes it very attractive for students. It serves as a remote experimental environment to exploit various control techniques and thereby to provide comparative study. At this time, the students can explore several control algorithms for various system configurations, including state-feedback, PID and fuzzy controllers.

For realistic presentation of the experiment via the Internet, several visualization methods were used. The students choose from video- and audiostreams, 3D animated graphics or conventional curve plotting, depending on the speed of their data connection to our lab. Only standard Web-browsers (optional with VRML Plug-In and Java Media Framework) are required as an interface to the laboratory.

Main components of our virtual laboratory and their interconnections are depicted in Fig. 1.

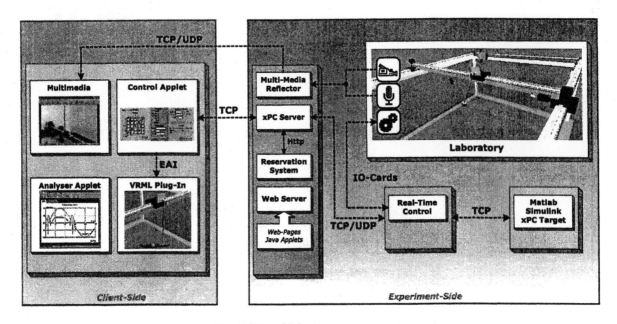

Fig. 1. Virtual laboratory components

Software components and tools used in the lab can be organized into three groups:

- Components for experiment set-up (control algorithms design, procedures set-up, real-time control of the pendulum/crane system) – Matlab/Simulink/xPC Target
- Server components interconnecting clients with the experiment – xPC Server communicating with the reservation system, Multimedia Reflector, Web Server
- Client components for remote lab operation and experiment visualization – a Control Applet, Java Media Framework, an Analyser Applet, and a VRML Plug-In

3. EXPERIMENT SET-UP

The essential part of the laboratory, the inverted pendulum/gantry crane system, is shown in detail in Fig. 2. It is a triaxial system, which consists of the gantry movable along an x-axis and pendulum/crane system also movable along y-axis placed on it. The DC motors drive both the gantry and the cart. For control purpose, a standard PC with an input/output cards is used. These cards converts the signals form the sensors (optical encoders to measuring of the axes positions; contactless proximity sensors) and control the axes drives.

Fig. 2. Inverted pendulum/gantry crane system

The control algorithms were designed and implemented by the Matlab/Simulink software package (Release 12) with Real-Time Workshop and xPC Target Toolbox. These tools are able to automatically generate stand-alone real-time applications from the Simulink models that run on the so-called *Target PC*, while their development is done on the host computer (*Host PC*). The Target PC is based on a special real-time operating system (Real-Time Kernel) and it communicates with the Host PC through serial link or network connection. Through this connection it is possible to operate the Target PC (load/start/stop application, change algorithm parameters, trace signals) completely by the Host PC (from the Matlab command line or directly from the Simulink model). The major advantages of this experimentation platform are: short development times for the control algorithms, only standard PC-hardware requested for the Target PC (486/8MB PC is often suitable minimum, we used a Pentium MMX/200MHz/32MB), and high stability and reliability of the Target PC, since it works with an optimised operating system (reduces to necessary functionality).

4. SERVER COMPONENTS

Server components control the Internet access to the lab and the experiment flow (xPC Server with reservation system) and exchange the data between the user and the lab.

4.1 xPC Server

Even though the Target PC is fully controllable from the Host PC, this feasibility was used only in the development and testing phase of the lab. For the teleoperation of the lab, some type of remote control of the Target PC via Internet is necessary. With the xPC Target Toolbox, a simple interface to the xPC Target through Web-browser is provided too. Since it was not possible to implement all required functionality with this interface, we decided to use C-API functions from the supplied library *xpcapi.dll* to build our own server application to manage the Target PC, called *xPC Server*. These functions serve to establish a basic control for the Target PC (restart, opening/closing of the communication port), setting of any real-time applications and control of its functional flow (load/unload, start/stop, stop time and sample time setting), adjustment of the control algorithms and signal tracing. On the other side, the xPC Server communicates with the client components, enables the control of the lab cameras and supervises the access to the lab.

One of the main advantages of the realised xPC Server is its easy adaptability to new experiments and other lab equipment. Namely, every experiment is implemented by a Simulink model. This model contains the particular control algorithm as well as some special blocks that are required for operation with the xPC Server. These blocks are:

- Blocks for the initialisation of the experimental plant
- Blocks for turning on/off of the control algorithm
- Blocks for acquisition and transmitting of the experimental data
- Block for the camera control

However, all these requirements are easy to accomplish, as the necessary blocks and subsystems are for the disposal in advance.

The second condition, under which is the experiment compatible with the xPC Server, is the initialisation file created to the compiled Simulink model. This file includes the information about the model, the list of the adjustable parameters with their boundary values, the numbers of the captured signals, etc. The xPC Sever reads this file before loading the model on the xPC Target and sets the parameters of the communication with the client accordingly.

In the xPC Server, two techniques for data logging are implemented. For data acquisition during the execution of the lab experiment, the fast (but unreliable) UDP protocol is used. This type of communication is supported directly by the xPC Target Toolbox. Therefore, only dedicated blocks must be added into the Simulink model with the tested algorithm. The xPC Target sends then during running of the application the data through its UDP port to the xPC Server, which transfers them further to the client. In this manner received data are used on the client-side for the animation of the 3D-model and on-line display.

For any more precise analysis of the experiment, another method for data storage is used. Since the data for analysis must be uniformly sampled, we used special blocks from the xPC Target Toolbox in Simulink models – *xPC Scopes*. The data are logged for desired time on the xPC Target at first and then send to xPC Server all at once. On the server side, these data are saved in files, which are available for download by the student.

The last function of the xPC Server is the operation of the cameras in the lab. In our lab, several cameras are placed that intervene a view on the experiment form various viewing angles. The user can teleoperate the cameras (rotating and zooming) and thereby adjust them according to his needs. Besides the pictures from the camera, the xPC Target screen is transmitted to the client, who can thereby observe its operating condition.

4.2 Multimedia Reflector

The *Multimedia Reflector* (MMR) is a server component developed during the LearNet project (Ruhr-Universität Bochum). It is designed for real-time processing of video-, audio- and data-streams from the laboratory and for their live transmission (synchronously) to the client. MMR simultaneously transmits streams to several clients with various transfer rates.

In addition to remote experimentation via Internet, MMR can be used during any conventional training, too. Here the students can observe the experiment on several workstations (e.g. for large groups, for small experimental set-up's). The MMR server is based on Java Media Framework (JMF) functions provided by Sun.

4.3 Reservation System

A new reservation system for distributed labs has been developed during this project and it is now used for booking of time quota with respect to each experiment in charge. This system is accessible through Web-browser interface and offers information about occupation of particular experiment. The sample reservation interface is presented in Fig. 3.

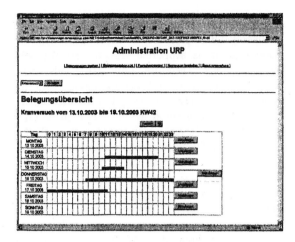

Fig. 3. Interface to the reservation system

Before accessing the lab, every user has to identify by his user name and password that are apportioned to him by the reservation system. These data are checked by the xPC Server, which communicates with the reservation system server via Http protocol. The answer of the reservation system is an XML-document with allowed duration of the experimentation and acknowledgement of the access to the lab. The xPC Server either executes this checkout in uniform time intervals during experimentation in order to assure the access to the laboratory for the next client.

5. CLIENT COMPONENTS

Client components are dedicated to experimental flow control and data visualisation. The communication with all servers involved is based on the TCP protocol and can be handled by standard Web-Browsers. All client components are implemented as JAVA-applets. Thereby, a comfortable and easy to operate user interface is created. The main requirement on the clients side can thus be reduces standard Web-browser without additional software.

Because of several types of available web-browsers and varied transfer rate of the Internet connection, a number of different equipped interfaces to the lab (full version for the students with the fast connection, the version with only VRML-model animation for the clients with slow connection, etc.) were designed. The Web-interface to the gantry crane controlled by a fuzzy controller is shown in Fig. 4.

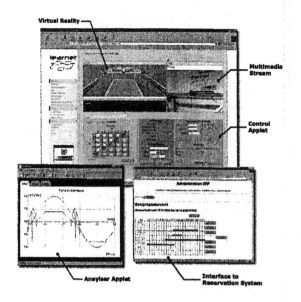

Fig. 4. Web-browser with user interface to the lab

5.1 Control Applet

The main component for teleoperation of the virtual lab is the *Control Applet*. A version for gantry crane control is shown in Fig. 5. The tasks performed by the Control Applet are:

- Verification of the access to the lab
- Opening/closing connection with the xPC Server
- Transmitting the commands to the server in defined protocol (start/stop application, get/set parameters, etc.)
- Displaying experiment state (elapsed time, server response, etc.) and signal values
- Controlling of the communication status and user activity (when no activity occurs, it aborts the experiment and closes the communication)
- Communication with any VRML Plug-In

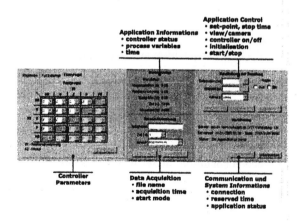

Fig. 5. Control Applet

5.2 VRML Plug-In

A VRML Plug-In is used for visualisation of the experiment by means of its virtual 3D model. This

144

option was used because it allows realistic visualization of the lab. On the other hand, it works even with low-bandwidth Internet connections, because of the small amount of transmitted data. Furthermore, virtual reality model enables to visualise miscellaneous physical and process quantities like acting forces, set-points, control signals, etc., and to interactively handle the lab by various switches and sliders. The animation of the virtual scene is implemented by using the External Authoring Interface (EAI).

The 3D model of the pendulum/crane system is shown in Fig. 6. In this model, the user can change the desired position of the crane by so-called touch sensor.

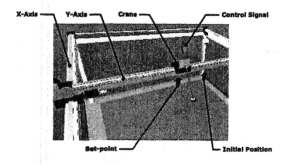

Fig. 6. VRML 3D model of the pendulum/gantry crane system

5.3 Analyser Applet

A special '*Analyser Applet*' has been developed for the on-line experiment evaluation; it is useable for plotting of the experimental data saved on the server. The data are received by the 'Control Applet' from the xPC Server and saved in the user's local file system. The user can plot several curves, zoom-in/out the graph, etc. This applet is shown in Fig. 7.

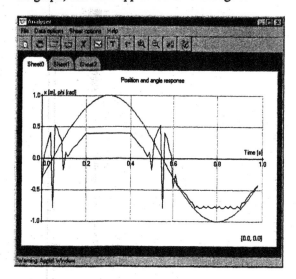

Fig. 7. Analyser Applet

5.4 Multimedia Applet

An additional 'Multimedia Applet' (Ruhr-Uni-Bochum: in combination with MMR server) displays streaming video from the virtual lab, plays sound and roughly plots the process data. It is a hybrid Java-class that can be started as Java applet or as stand-alone application. It exploits JMF, which must be installed on the client computer. On the start-up, the user can chose from the set of available media (video, sound, data) as well as transfer rate for the video stream. This 'Multimedia Applet' receives the data directly from the MMR server. Examples of video streams from several cameras located in our lab as well as transmitted xPC Target screen are given in Fig. 8.

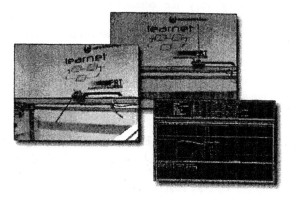

Fig. 8. Video streams from the lab

6. LABORATORY UTILIZATION

As the FernUniversität in Hagen is the only distance teaching university, such virtual labs play significant role in its educational concept, particularly in the engineering-oriented fields of study.

The presented laboratory provides a wide range of executable experiments with various complexity and it can be used as the test platform for verification of many types of the control strategies. The spectrum of the experiments ranges from the position/speed control of the single axis to the control of the pendulum (alternatively 2D-pendulum) in its stable/unstable position by several types of controllers (PID, state-feedback, fuzzy controller, etc.).

The laboratory was completed during the summer term of the last year and the first students were testing the lab also in the last summer and winter term. The tests were performed through both LAN and modem connection with satisfactory results. At the present time, we offer the exercise in the gantry crane configuration as the part of the several courses from the control theory and mechatronic.

CONCLUSION

This paper describes the concept and implementation of our new 'virtual laboratory' for the inverted pendulum/gantry crane control.

The components used in the lab as well as on the client side have been discussed here. The novel architecture of the virtual laboratory based on utilization of the xPC Target Toolbox from Mathworks was designed. For the communication with the user and operating of the xPC Target, the xPC Server was developed. The presented structure of the lab together with its components provides the framework for future labs.

The technical structure of the lab allows further addition of new control algorithms/problems to solve and thereby vary the education process for the students.

On the client side, only minor demands are made. Moreover, the students can choose among several methods for experiment presentation.

The first experimentation proved the functionality of the lab and its applicability by education process.

ACKNOWLEDGEMENT

This work was supported by a grant of the German Federal Ministry for Education and Research under Grant No. 08 NM 101D. The authors are responsible for the content of this paper.

REFERENCES

Röhrig C., Jochheim A. (2000). Java-based Framework for Remote Access to Laboratory Experiments, In: IFAC/IEEE Symposium on Advances in Control Education, ACE 2000, Gold Coast, Australia

Bischoff A., Röhrig C. (2001). A Multiuser Environment for Remote Experimentation in Control Education, IFAC Workshop on Internet Based Control Education, IBCE'01, Madrid, Spain

Real Systems in Virtual Lab,
http://rsvl.fernuni-hagen.de

SPS-Training (joint project with Siemens AG),
http://prt.fernuni-hagen.de/sps/spsrail

BMBF (German Federal Ministry for Education and Research) Project „Learnet",
http://www.learnet.de

Copyright © 2004. Published by Elsevier Ltd
on behalf of IFAC

ELSEVIER
IFAC
PUBLICATIONS
www.elsevier.com/locate/ifac

COLLABORATING WITH A DYNAMICALLY AUTONOMOUS COGNITIVE ROBOT

Donald Sofge[1], Dennis Perzanowski[1], Marjorie Skubic[2], J. Gregory Trafton[1],
Magdalena Bugajska[1], Derek Brock[1], Nicholas Cassimatis[1],
William Adams[1], Alan Schultz[1]

[1]*Navy Center for Applied Research in Artificial Intelligence*
Naval Research Laboratory
Washington, DC 20375

[2]*Electrical and Computer Engineering Department*
University of Missouri-Columbia
Columbia, MO 65211

Abstract: Achieving effective cooperation between robots and humans requires the design of a natural, human-centric interface and autonomous robot behaviors whereby a human can communicate and interact with a robot almost as efficiently as he/she would with another human. In this interaction the human acts as a supervisor and collaborator with the robot, rather than strictly as a teleoperator. This requires that a number of capabilities be built into the robot and robot interface, including voice recognition, natural language and gesture understanding, a cognitive model, and behaviors supporting dynamic autonomy. This paper describes the design and implementation of such a system.

Keywords: Autonomous Mobile Robots, Cognitive Systems, Co-operative Control, Speech Recognition, Natural Language, Human-Machine Interface.

1. INTRODUCTION

Effective collaboration between robots and humans in accomplishing complex tasks requires a number of capabilities not often found in present-day service robots. These include voice recognition and natural language understanding, recognition of human gestures (such as pointing to objects), the ability to reason about objects in space as humans do, and built-in behaviors for sequencing and executing tasks requiring various levels of control by and interaction with a human supervisor. One of the key challenges in implementing dynamically autonomous behaviors in mobile robots is achieving a truly human-centric interface so that human operators can interact with the robots as naturally as they would with another human.

In this effort we facilitate the natural interaction between humans and robots through use of a multimodal interface (Perzanowski, *et al.*, 2000). We define "human-centric" as focusing on the needs and natural modes of interaction of the human rather than the robot. A key feature of this interface is the use of multiple overlapping modes of communication between the operator and the robot. These overlapping (and sometimes redundant) modes of communication provide the operator with a natural interface to the system, allowing the operator to choose the mode of communication most comfortable to him/her given the current task, situation and environmental conditions.

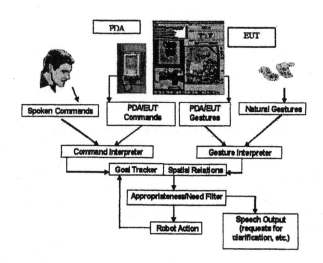

Fig. 1. Human-Centric Multimodal Interface

Redundancy in communications is also beneficial in case of subsystem failures, and may also serve to help resolve ambiguity, such as when a command is communicated in more than one way to the robot (e.g. verbally and through a gesture).

Use of a cognitive model aboard the robot further facilitates better communication and interaction between the human and the robot through use of a common representational framework for the environment and objects within it, processing of sensor information, and joint problem solving involving both humans and robots.

2. DYNAMIC AUTONOMY

Dynamic autonomy allows the robot to dynamically adjust its behaviors depending upon and appropriate to the task(s) at hand (Perzanowski, *et al.*, 1999). The human operator is able to interact with the robot in a human-centric manner by providing verbal commands and gestures to the robot to perform tasks requiring varying levels of human interaction. Some circumstances may require very fine-grained level operator control, while others require less precision.

Dynamic autonomy used in mobile robots provides a more flexible and operator-friendly interface and makes the robots more versatile.

We support dynamic autonomy in our system through a number of robot behaviors of varying complexity including collision-free navigation, path following, exploration, automatic prioritization of multiple command directives, and feedback from the robot to the operator. Feedback is provided by voice synthesis and through text strings requesting clarification if the robot isn't able to understand the command(s). Operation of the natural language

interface with the gesture interpretation process and other command input modes is discussed in greater detail in the section describing the robot's integrated goal-driven architecture. Figure 1 shows our human-centric multimodal interface for autonomous mobile robots.

Using this interface, commands may be communicated to the robot in a variety of human-centric ways including verbally, through touch (on a PDA, touch tablet or keyboard), and by gesturing with hands and arms. The robot passes a variety of information back to the operator such as sensor readings, video, navigation maps built using its array of on-board sensors, linguistic descriptions of its environment (Skubic *et al.*, 2002, 2004), and the status of commands previously sent to it. The interface is easily adapted to a variety of computing devices and platforms selected for the operational environment, including PDAs, End-User-Terminal (EUT), and tablet PCs, as shown in Fig. 1, with display and information scaled appropriately.

3. UNDERSTANDING GESTURES

One of the key modes of interaction with the robot is through the use of gestures. Several types of gestural interfaces have been developed and used in the past. For example, one gestural interface uses stylized gestures of arm and hand configurations (Kortenkamp, *et al.*, 1996), while another is limited to the use of gestural strokes on a PDA display ("synthetic" gestures) (Fong, *et al.*, 2000). In our system we combine both of these approaches, allowing both stylized and synthetic gestures. Our stylized gesture interface utilizes a structured-light rangefinder to detect the positions of the hands over several consecutive frames to generate trajectories for the gesture command. The structured-light rangefinder emits a horizontal plane of laser light. A camera mounted on the robot just above the laser is fitted with an optical filter which is tuned to the frequency of the laser. The camera registers the reflection of the laser light off of objects in the room and generates a depth map (XY) based upon location and pixel intensity. The data points for bright pixels (indicating closeness to the robot) are clustered. If a cluster is significantly closer to the robot than background scenery, it is interpreted as a hand. Hand locations are stored from several consecutive frames, and the positions of the hands are used to generate trajectories for the gesture command. Each trajectory is analyzed to determine if it represents a valid gesture. The command corresponding to the matched gesture is then queued so that the multimodal interface, upon receiving another command, can retrieve the gesture from the gesture queue and combine it with the verbal command in the command interpretation system.

4. NATURAL LANGUAGE INTERFACE

Our natural language interface combines a commercial speech recognition front-end with an in-house developed deep parsing system (Perzanowski, *et al.*, 1999). ViaVoice™ is used to translate the speech signal into text, which is then passed to our natural language understanding system, NAUTILUS (Wauchope, 1994), to produce both syntactic and semantic interpretations. The semantic interpretation, interpreted gestures from the vision system, and command inputs from the computer or other interfaces are compared, matched and resolved in the command interpretation system (Figure 1).

Using our multimodal interface the human user can interact with the robot using both natural language and gestures. The semantic interpretation is linked, where necessary, to gesture information via the *Gesture Interpreter*, *Goal Tracker*, *Spatial Relations* component, and *Appropriateness/Need Filter*, and an appropriate robot action or response results.

For example, the human user can ask the robot "How many objects do you see?" ViaVoice™ analyzes the speech signal, producing a text string. NAUTILUS parses the string and produces a representation something like the following, simplified here for expository purposes.

```
(ASKWH                                  (1)
    (MANY N3 (:CLASS OBJECT) PLURAL)
    (PRESENT #:V7791
    (:CLASS P-SEE)
    (:AGENT (PRON N1 (:CLASS SYSTEM)YOU))
        (:THEME N3)))
```

The parsed text string is mapped into a kind of semantic representation, as in (1), in which the various verbs or predicates of an utterance (e.g. *see*) are mapped into corresponding semantic classes (*p-see*) that have particular argument structures (*agent, theme*). For example *you* is the agent of the *p-see* class of verbs in this domain and a class of plural objects is the theme of this verbal class, represented as *N3*—a kind of co-indexed trace element in the theme slot of the predicate, since this element is fronted in English wh-questions. If the spoken utterance requires a gesture for disambiguation, as in for example the sentence "Look over there," the gesture components obtain and send the appropriate gesture to the *Goal Tracker* and *Spatial Relations* components which combine linguistic and gesture information.

5. SPATIAL REASONING

Spatial reasoning is an important element of a human-centric interface because humans often think in terms of relative spatial positions, and use such relational linguistic terminology naturally in communicating with one another. Our spatial reasoning component builds upon an existing framework of natural language understanding with semantic interpretation (Skubic, *et al.*, 2002), and utilizes on-board sensors for detecting objects and map-building through use of evidence grids.

Understanding spatial linguistic terms allows for more efficient and natural control of a dynamically autonomous mobile robot. For example, we may want to give the robot a command such as "Go down the road 100 meters, turn left behind the building on your right, and proceed ahead 30 meters until you are in front of the tower. Then go into active surveillance mode." Or, in an office setting, "Go between the table and the chair, through the doorway, and down the hall to the left 20 meters."

The spatial reasoning component of the multimodal interface allows the robot to provide feedback to the human operator using natural spatial terminology that is easy for the human to understand. The human is able to query the robot about the relative spatial positions of objects in the environment in the same simple and natural language, and the robot is able to respond using spatial terms. This is demonstrated in the following dialogue (2):

Human: "How many objects do you see?"   (2)
Robot: "I see three objects."
Human: "Where are they located?"
Robot: "Object A is five meters in front of me.
    Object B is ten meters in front of me
    and to my right.
    Object C is twenty meters to my left."

This natural spatial language is used to disambiguate spatial references by both humans and robots

(Skubic, *et al.*, 2002). It provides a common interpretation for location expressions, such as "left" and "right," as well as other relative directions. For example, if the human commands the robot, "Turn left," the robot must understand whose left is being referred to, the human's or the robot's. Use of spatial language between humans and robots is currently under investigation by our group at NRL through human-factors experiments (Perzanowski, *et al.*, 2003) where novice users provide instructions to the robot for performing various tasks where spatial referencing is required. This work will result in development of a common language for spatial referencing geared to the needs and expectations of untrained and non-expert operators. This common spatial language will be incorporated into the multimodal interface.

6. EMBODIED COGNITION

Achieving effective collaboration between humans and robots is greatly facilitated by the use of cognitive models on-board the robots. We refer to this as *embodied cognition*. Embodied cognition, using cognitive models of human performance to augment a robot's reasoning capabilities, facilitates human-robot interaction in two ways. First, the more a robot behaves like a human being, the easier it will be for humans to predict and understand its behavior and interact with it. Second, if humans and robots share at least some of their representational structure, communication among them will be much easier. For example, in tasks requiring direction generation, humans naturally use qualitative spatial relationships (Miller and Johnson-Laird, 1976; Tversky, 1993) such as "left," "up," "east," and "north." Interaction with a robot capable of manipulating the same representation instead of traditional real number matrices would be more natural and efficient. In (Bugajska, *et al.*, 2002) and (Trafton, *et al.*, 2003) we used cognitive models of human performance of the task to augment the capabilities of robotic systems.

We use two cognitive architectures based on human cognition for certain high-level control mechanisms in our robots. These cognitive architectures are ACT-R (Anderson and Lebiere, 1998) and Polyscheme (Cassimatis, 2002).

ACT-R is one of the most prominent cognitive architectures to have emerged in the past two decades as a result of the information processing revolution in the cognitive sciences. Also called a unified theory of cognition, ACT-R is a relatively complete theory about the structure of human cognition that strives to account for the full range of cognitive behavior with a single, coherent set of mechanisms. Its chief computational claims are: first, that cognition functions at two levels, one symbolic and the other sub-symbolic; second, that

symbolic memory has two components, one procedural and the other declarative; and third, that the sub-symbolic performance of memory is an environmentally optimized response to the statistical structure of the environment. These theoretical claims are implemented as a production-system modeling environment. The theory has been successfully used to account for human performance data in a wide variety of domains including using memory for goals (Altmann and Trafton, 2002), human computer interaction (Anderson, *et al.*, 2002), and scientific discovery (Schunn and Anderson, 1998). We use ACT-R to create cognitively plausible models of appropriate tasks for the robots to perform.

Second, we use Cassimatis' Polyscheme architecture (Cassimatis, 2002) for spatial, temporal and physical reasoning. The Polyscheme cognitive architecture enables multiple representations and algorithms (including ACT-R models), encapsulated in "specialists," to be integrated into inference about a situation. We use an updated version of the Polyscheme implementation of a physical reasoner to help keep track of the robot's physical environment.

6.1 Perspective-Taking

One feature of human cognition that facilitates natural human-robot interaction is "perspective-taking". In order to understand utterances such as "the wrench on my left," the robot must be able to reason from the perspective of the speaker to resolve the meaning of "my left". We use the Polyscheme architecture and ACT-R models to endow the robot with the ability to conceive of task-oriented goals and knowledge of another person. This allows the robot to more easily predict and explain its own behavior, as well as the behavior of others, making it a better partner in a collaborative activity.

Polyscheme has a simulation mechanism, called a "world" in which the robot is endowed with perspective-taking capabilities. Polyscheme allows the robot to reason about what it sees in its immediate environment from different perspectives. Using worlds, Polyscheme can simulate the perspective it would have at other times, different places and in other hypothetical worlds, and use its specialists to make inferences within those perspectives. Polyscheme uses reasoning algorithms such as counterfactual reasoning, backtracking search, truth-maintenance and stochastic simulation. We have created a specialist to reason about the perspective(s) of *other people*. This allows Polyscheme to predict and explain other people's behavior, using its perceptual, motor, procedural, memory, spatial and physical specialists from another person's perspective.

For both ACT-R/S and Polyscheme we have created models that can perform simple spatial perspective-taking tasks. There seem, however, to be advantages and disadvantages to both systems: For example, ACT-R/S has more difficulty doing large scale simulations, but has a large amount of historical cognitive plausibility (e.g., there have been a large number of empirical and psychological studies validating ACT-R/S), while Polyscheme has comparatively less of a cognitive history. Additionally, because the representations and operations of each system are a bit different, their behaviors are different and various tasks may be easier or more straightforward to model for one system than for another.

7. MOBILE ROBOT INTEGRATED GOAL-DRIVEN ARCHITECTURE

Figure 2 shows our mobile robot integrated goal-driven architecture. This architecture is organized by the integration and arbitration of goals presented though various interface modules. Outputs for speech recognition, natural language understanding, gesture interpretation, and other interface modules are cached; command prioritization and resolution are then performed.

Once goals are interpreted and resolved, they are passed to the *Path Planning and Navigation* routines, where they are integrated with low-level behaviors such as obstacle avoidance, exploration and path planning using the *Vector Field Histogram* (VFH) method (Borenstein and Koren, 1991). The architecture maintains both short-term and long-term maps (not shown in Figure 2), which are also important for several of the other processes such as *Spatial Reasoning*, *PDA Interface*, and *Robot GUI*.

8. CONCLUSIONS

This paper describes the design, implementation, and capabilities of a robotic system architecture for a robot which can be used to collaborate with a human. The capabilities required of the robot include voice recognition, natural language understanding, gesture recognition, spatial reasoning, and cognitive modeling with perspective-taking. These represent a small subset of potential capabilities humans utilize in collaborating with one another to perform a task in a complex environment. However, we barely scratch the surface of capabilities we might want to build into an intelligent, collaborative robot. We are currently performing human-subject experiments on cooperation in task performance — both with and without robots — in order to gain a better understanding of which capabilities are most critical aboard the robot.

Most of the capabilities described above have been successfully implemented on several robotic platforms and demonstrated. We are currently adding the cognitive architecture (both ACT-R and Polyscheme) and perspective-taking capabilities. Future work will focus on enhancing the cognitive models through expanded rulesets and cognitively plausible (in human terms) behaviors and reasoning mechanisms, and by adding learning capabilities to the models. Parts of this architecture are also being extended to several robots designed specifically for enhanced human interaction, namely NASA's humanoid robot Robonaut (Ambrose, et al., 2000) and MIT's clearly non-humanoid robot Kismet (Breazeal, 2003).

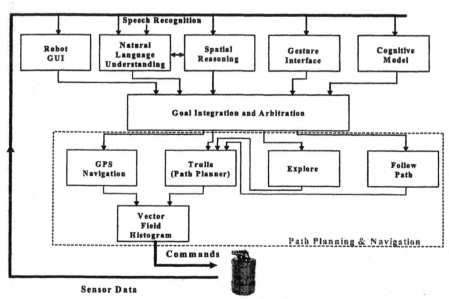

Fig. 2. Mobile Robot Integrated Architecture

ACKNOWLEDGEMENTS

Support for this effort was provided by the DARPA
IPTO Mobile Autonomous Robot Software
(DARPA MARS) Program. Thanks also to Sam
Blisard and Scott Thomas for their contributions to
this effort.

REFERENCES

Altmann, E. M. and J. G. Traftonm (2002). "An
activation-based model of memory for goals," In
Cognitive Science, 39-83.

Ambrose, R. O., H. Aldridge, R. S. Askew, R. R.
Burridge,W. Bluethmann, M. Diftler, C.
Lovchik, D. Magruder, F. Rehnmark (2000).
"Robonaut: NASA's space humanoid," IEEE
Intelligent Systems, *IEEE Intelligent Systems*,
vol. 15, no. 4 , pp. 57-63.

Anderson, J. R. and C. Lebiere (1998). *The atomic
components of thought*. Lawrence Erlbaum.

Anderson, J. R., M. Matessa, and C. Lebiere
(1997). "ACT-R: A theory of higher level
cognition and its relation to visual attention," In
Human-Computer Interaction, 12 (4), 439-462),
ASME Press, 763-768.

Borenstein, J. and Y. Koren (1991). "The Vector
Field Histogram – Fast Obstacle Avoidance for
Mobile Robots," *IEEE Transactions on Robots
and Automation*, IEEE: New York.

Breazeal, C. (2003). *"Towards sociable robots,"*
Robotics and Autonomous Systems, vol. 42, no.
3-4.

Bugajska, M., A. Schultz, T. J. Trafton, M. Taylor,
and F. Mintz (2002). "A Hybrid Cognitive-
Reactive Multi-Agent Controller," In
*Proceedings of 2002 IEEE/RSJ International
Conference on Intelligent Robots and Systems
(IROS-2002)*, EPFL, Switzerland.

Cassimatis., N. L. (2002). "Polyscheme: A
cognitive architecture for integrating multiple
representation and inference schemes," PhD
dissertation, MIT Media Laboratory.

Fong, T. W., Conti, F., Grange, S. and Baur, C.
(2000), "Novel Interfaces for Remote Driving:
Gesture, haptic, and PDA," *SPIE 4195-33, SPIE
Telemanipulator and Telepresence Technologies
VII*, Boston, MA.

Kortenkamp, D., E. Huber, and P. Bonasso, P.
(1996). "Recognizing and Interpreting Gestures
on a Mobile Robot," In *Proceedings of AAAI*.

Miller, G. A., and P. H. Johnson-Laird (1976).
Language and Perception. Harvard University
Press.

Perzanowski, D., A. Schultz, W. Adams, and E.
Marsh (1999). "Goal Tracking in a Natural
Language Interface: Towards Achieving
Adjustable Autonomy," In *Proceedings 1999
IEEE International Symposium on
Computational Intelligence in Robotics and
Automation*, Monterey, CA.

Perzanowski, D., W. Adams, A. Schultz, and E.
Marsh (2000). "Towards Seamless Integration in
a Multimodal Interface," In *Proceedings 2000
Workshop Interactive Robotics and
Entertainment*, pages 3-9, Menlo Park, CA.

Perzanowski, D., D. Brock, S. Blisard, W. Adams,
M. Bugajska, A. Schultz, G. Trafton, M. Skubic,
(2003), "Finding the FOO: A Pilot Study for a
Multimodal Interface," In *Proceedings of the
IEEE Systems, Man, and Cybernetics
Conference*, Washington, DC.

Schunn, C. D. and J. R. Anderson (1998).
"Scientific discovery," In J. R. Anderson, and C.
Lebiere (Eds.), *Atomic Components of Thought*.
Lawrence Erlbaum.

Skubic, M., D. Perzanowski, A. Schultz, and W.
Adams (2002). "Using Spatial Language in a
Human-Robot Dialog," In *Proceedings 2002
IEEE Conference on Robotics and Automation*,
IEEE.

Skubic, M., Perzanowski, D., Blisard, S., Schultz,
A., Adams, W., Bugajska, M. and Brock, D.
(2004) "Spatial Language for Human-Robot
Dialogs," In *IEEE Transactions on SMC Part C*,
in press.

Trafton., J. G., A. Schultz, D. Perzanowski, W.
Adams, M. Bugajska, N. L. Cassimatis, and D.
Brock (2003). "Children and robots learning to
play hide and seek," In *Proceedings of the IJCAI
Workshop on Cognitive Modeling of Agents and
Multi-Agent Interactions*, Acapulco, Mexico.

Tversky, B. (1993). "Cognitive maps, cognitive
collages, and spatial mental model," In A. U.
Frank and I. Campari (Eds.), *Spatial information
theory: Theoretical basis for GIS*, Springer-
Verlag.

Wauchope, K. (1994). *"Eucalyptus: Integrating
Natural Language Input with a Graphical User
Interface,"* Naval Research Laboratory Technical
Report NRL/FR/5510-94-9711, Washington,
DC.

Copyright © IFAC Telematics Applications in Automation and Robotics, Espoo, Finland, 2004

ELSEVIER
IFAC
PUBLICATIONS
www.elsevier.com/locate/ifac

DEVELOPMENTS OF TRACK LOCOMOTION SYSTEMS FOR PLANETARY MOBILE ROBOTS

A. Bogatchev [(1)], **V. Koutcherenko** [(1)], **M. Malenkov** [(2)], **S. Matrossov** [(1)]

[(1)] Science & Technology Rover Company Ltd (RCL)
[(2)] J.-St. Co. Russian Mobile Vehicle Engineering Institute (VNIITRANSMASH)
2 Zarechnaja Street, 198323, St. Petersburg (Russia)
rover@peterlink.ru; rcl@rcl.spb.su
www.rovercompany.spb.ru

Abstract

Since 1962 the J.-St. Co. VNIITRANSMASH and then Science & Technology Co. Ltd. (RCL) carried works on development of locomotion systems for planetary rovers. One of directions of these works is development of track locomotion systems. This paper considers some of track locomotion systems developed within the framework of the Lunokhod 1 and 2 program, MARSNET and RoSa-2 projects. *Copyright © 2004 IFAC*

1. Introduction

One of main problems decided while creating mobile robots including planetary rovers is to ensure its high cross-country ability on soft soils and the rough surface. It is especially important for Mars rover, where distance precludes the teleoperated methods that are practical at the Moon. Besides the on-board intelligence is required for Mars. Higher cross-country ability promotes reduction requirements to the guidance-control system.

Development of mobility systems is performed basically in two main directions: wheeled and tracks ones. Wheeled chassis especially multiwheeled are the most widespread as they have greater reliability of a mover [1,2]. At the same time a track chassis have high traction and cohesion characteristics that is important when climbing friable soil slopes [4]

Track movers in USA designed and built by the Bendix Corporation and JPL [3]. The Bendix's vehicle has an articulated body and mover in the form of four track modules. Cross-country of the mover was insufficient. Beginning from the sixties JPL serious investigated various aspects of robotics including Mars rovers.

One of models of Mars rover configuration, used as Bendix's rover, the four-track-module mover. But the use of articulated suspension arms for each module provided high mobility of the model.

Works on development of track locomotion systems for planetary rovers in the Soviet Union and then in Russia are carried out at VNIITRANSMASH, where the Lunokhod's 1 and 2 chassis was developed and built, and Kemurdjian STC Rover Co. Ltd (RCL). Some their developments are consider below.

2. Track locomotion systems

While creating the chassis for Lunokhod 1 and Lunokhod 2 the track mover was studied side by side with wheeled movers. Fig. 1 shows one of the first versions of chassis for Lunokhods with the two-track mover. Each track has the individual drive (motor and gearhead) and wheels have individual torsion bar suspension. A control system provided moving with different speeds and turn at the expense of different speeds of tracks on sides when the mock-up moves and turn on the spot when the tracks rotates in the opposite directions (tractor-type turn).

Fig.1. Fist Mock-up of "Soviet Lunokhod" with the two-track mover
(1965, mass-60 kg,; speed-0,5 km/h; dimension- 1,5 x 1,2 x 0.4m ; soil slope- 25 dgr)

The main shortcoming of the two-track mover was lamming the track as result of ingress of rocks into engagement of the track with rotating parts. This shortcoming is redoubled because of limited power for the rover motion in the budget of power inputs for planetary rovers. Other shortcomings are insufficient smoothness of running the rover on the rough surface, while surmounting projecting obstacles and heaping up soil when turning especially on the spot.

One of directions of improvement of the track mover as applied to transport robots including planetary rovers is dismemberment of the mover on a number unified modules [1]. Numerous projects provide for creation of mobile rovers on the basis of four-track mover. As example of such a mover Fig.2 shows a running mock-up developed by VNIITRANSMASH (Designer: Dr. A.Egorov) as applied to a large Mars rover. Its main parameters are:

-mass, kg	367
-mover base, m	1.7 m
-mover width, m	1.9 m
-ground clearance, m	0.55
-obstacle surmounted, m, max. (height)	0.4
-friable soil slope climbed, dgr, max	29
-height of the centre of gravity, m	0.7

The mover has four unified track modules and each of them is equipped with electromechanical drive and the individual independent torsion-bar suspension. Fig.3 and Fig 4. shows mobility for tracking mode and wheeled mode

obstacles   on a sandy slope   turning

Fig 2 .Tests of Running mock-up with the four-track mover
(barchans at the desert Kara Kum)

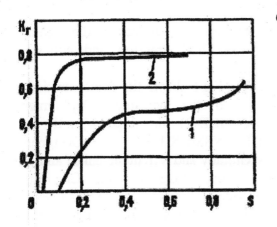

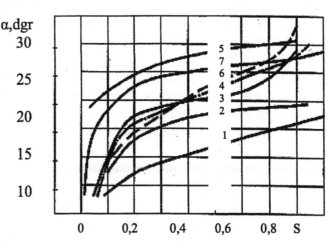

Fig 3. Graf of mobility (cross-country ability) depending on types of propulsive devices
1 – wheeled mode
2 - tracking mode

Fig 4. A plot of the traction coefficient (kt=tgα) and steepness of friable soil slopes (α) as a function of the slipping coefficient (s) for different movers.
1-wheelled mover 6x6 with rigid wheels
2-wheeled mover 6x6 with metal-elastic net wheels
3-wheeled mover 4x4 with metal-net types on rigid wheels (LRV)
4-wheel-walking mover with rigid wheels (interrupted "gait")
5 the same mover (interrupted "gait")
6- four -track mover
7-ski-walking mover

Fig.5 shows the separate track module. It contains all elements necessary for track-mover drive, its cushioning, and adjusting the track tension. Essential difference of the track module from the traditional track design is presence of high volumetric grousers with the wire netting surface to decrease ingress of soil into the module. The grousers are mounted on the track metal belts. Such decision excludes necessity to use supporting rollers and provides sufficiently high load-carrying capacity.

Fig. 5 Track module.

When turning on the spot high grousers do not prevent ingress of soil inside the module. So the module internal volume, where the drive, supports for suspension and other elements are placed, is closed from two sides with lateral shields.

The module has double cushioning. The basic torsion bar is connected with the upper end of the suspension lever and is fixed in the bracket. The additional torsion bar is connected with the lower end of the suspension lever and is fixed on the drive body. The elastic characteristic of the additional torsion bar can be adjusted.

Features enumerated of the track module design increase essentially its

serviceability and, as result, the vehicle mover has high traction and cohesion properties.

The Fig.6 shows the demonstrational mock-up of the Instrument Deployment Device (IDD-2) developed jointly by VNIITRANSMASH (Designers: Dr.V.Gromov and V.Kutcherenko) and the Max Plank Institute for Chemistry (Dr.R.Rider, Germany) for the MARSNET project (1994). This micro rover consists of three modules connected with each other with levers having drives in each module. The extreme modules are equipped with tracks set in motion by means of special drives. The track is a ring-form metal belt with grousers.

Fig. 6. Instrument Deployment Device IDD-2.

The track locomotion system provided a travel speed up to 0.94 m/s. The middle module, bearing the TV-camera and scientific equipment can be exactly positioned relatively of the object investigated. It also can be used to realize "walking" mode of moving the rover under complex terrain conditions. In this case the modules rotate relatively each other with different sequence depending on conditions of moving (soil, relief). As result the mini rover can surmount obstacles with height exceeding some times a height of the module.

Main parameters of IDD-2 are the following:

-mass, kg   1.5
-overall dimensions, mm   200x200x x50
obstacles surmounted:
-friable sole slope, dgr   29

-slope with a carpet path, dgr up to   65
-bench, mm up to   300 (height)
-crack, mm up to   300 (width)

The micro rover IDD-2 was a prototype for creating the micro robot RoSa-2 ("Micro robot for scientific applications 2") within the framework of the ESA's Technological Research and Development program (1999-2001) (Fig.7) [5]. The locomotion system developed and built by RCL (Designers: V.Kucherenko, A.Bogatchev, S.Vladykin and Project Manager- S.Matrossov) under contract with ESA is intended to transport a payload as well as to move the drilling unit (the part of the payload), which was developed by Helsinki university of technology HUT (prof. A.Halme, J.Suomela, J.Saaren) in co-operation with

Space systems Finland (M. Antilla) and VTT Automation (T.Ylikorpi).

The locomotion system consists of the body, two tracks, mechanism for vertical displacement of the drilling device. The track drives are placed in the track units and contain the motor, gear head and horizontal screw which engages into mesh with ball-bearing mounted on track. The robot is commanded and supplied with power from lander via a tether.

The locomotion system has the following main characteristics:
-maximum motion speed
on the horizontal surface, m/h   60
-mass of the chassis/payload
mass, kg   5/7
-overall dimensions in stowed
position, dgr   400 x 400 x 110
-slope climbed (with minimum ground clearance), dgr up to   30
-obstacle surmounted (with minimum ground clearance)   60 (height)

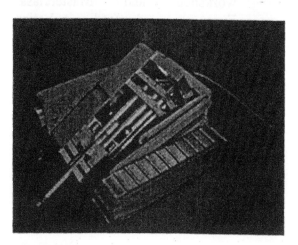

a) Position for operation;

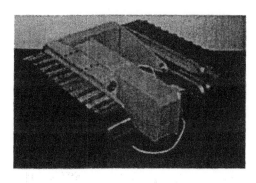

b) transport position without drill and tether

c) track

Fig 7. Micro robot RoSa-2

3. Conclusion

The track locomotion systems considered can provide a high cross-country ability and maneuverability on the complex terrain (soil, relief). They can be used for planetary rovers and Earth-based robots subject to their purpose. Accumulated experience allows perform further investigations and developments of track movers.

References.

Planetokhodi (Planetary rovers). Edited by A.L. Kemurdjian. 2nd revised and added edition. "Mashinostroenie", Moscow,1993 (in Russian).

1   Wheeled propulsive devices for mobile robots. Bogatchev A., Gromov V., Kucherenko V., Matrossov S. and et al., EUREL workshop and Masterclass. European advanced robotics systems development. Vol.2. Manchester UK, April, 2000.

2   Some details of the development of mobile platforms. Bogatchev A ,V. Gromov, M. Kucherenko V., Matrossov S , Malenkov, S. Fedoseev, et al. Proceedings of the international conference on field and service robotics. Canberra, Australia, 8-10 December, 1997.

3   James D. Burke. Past US studies and developments for planetary rovers. The international symposium on planetary mobile vehicles. Toulouse (France), September. 1992.

4   Joint RCL & HUT developments for mobile robot locomotion systems during 1995-2002. A. Bogatchev, V. Kutcherenko, S. Matrossov, S. Vladykin, V. Petriga, A. Halme, J. Suomela, I. Leppanen, S. Yionen., S. Salme 7th ESA workshop on advanced space technologies for robotics and automation, ESA/ESTEC. Noordwijk, The Netherlands, November 19-21, 2002.

Copyright © IFAC Telematics Applications in Automation
and Robotics, Espoo, Finland, 2004

RULE-BASED TRAVERSABILITY INDICES FOR MULTI-SCALE TERRAIN ASSESSMENT

Homayoun Seraji
NASA-Jet Propulsion Laboratory
California Institute of Technology
Pasadena, CA 91109, USA

Abstract

This paper presents novel measures of terrain traversability at three different scales of resolution; namely Local, Regional, and Global Traversability Indices. The Local Traversability Index is related by a set of linguistic rules to local obstacles and surface softness, measured by on-board sensors mounted on the robot. The rule-based Regional Traversability Index is computed from the terrain roughness and slope that are extracted from video images obtained by on-board cameras. The Global Traversability Index is obtained from the terrain topographic map, and is based on the natural or man-made surface features such as mountains and craters. Each traversability index is represented by four fuzzy sets with the linguistic labels {POOR, LOW, MODERATE, HIGH}, corresponding to surfaces that are unsafe, moderately-unsafe, moderately-safe, or safe for traversal, respectively [1]. Copyright © 2004 IFAC

1 Introduction

Exploration of planetary surfaces and operation in rough terrestrial terrain have been strong motivations for research in autonomous navigation of field mobile robots in recent years. These robots must cope with two fundamental problems. The first problem is to acquire and analyze the terrain quality information on-line and in real time, and to utilize it in conjunction with limited prior terrain imagery. The second problem is to deal with imprecision in sensor measurements and uncertainty in data interpretation inherent in sensing and perception of natural environments. With respect to these two fundamental problems, outdoor robot navigation defines a research topic that is more challenging than indoor robot navigation in structured and benign man-made environments.

Robust on-line terrain characterization and traversability assessment are clearly core research problems for autonomous field robot navigation. Two types of solutions have been proposed to date by researchers at CMU and JPL. In the CMU methods [1-6], the terrain traversability is computed along different candidate paths that correspond to different robot steering angles. The traversability of each path is determined mathematically by a weighted sum of the roll, pitch, and roughness of the grid-based map cells along that path, incorporating their certainty values [1]. The JPL rule-based approach [7-12] takes a sharp departure from analytical methods and is centered on the *Fuzzy Traversability Index*. This index is a novel concept that was first introduced in [7-8] as a simple *linguistic* measure for quantifying the suitability of the regional terrain for traversal by a mobile robot. This perceptual approach to terrain assessment is highly robust to measurement noises and interpretation errors because of the use of fuzzy sets in a linguistic rule-based system. This approach is analogous to the human judgment, reasoning, and decision-making regarding assessment and traversal of a natural terrain.

Earlier papers by JPL researchers [7-10] have focused on *regional* terrain characterization and traversability assessment. In this paper, the regional traversability index concept is extended to both *local*

---

[1]The research described in this paper was performed at the Jet Propulsion Laboratory, California Institute of Technology, under contract with the National Aeronautics and Space Administration.

and *global* terrain, to complement the regional measure. The paper is structured as follows. Sections 2-4 discuss terrain traversability analysis at local, regional, and global scales. The paper is concluded in Section 5 with a review of key features.

2 Local Traversability Analysis

For local traversability analysis, we focus on the terrain quality in close proximity of the mobile robot. Typically, this covers a distance of up to 0.5 meters away from the robot for a small-sized robot. Different measures of local terrain quality can be considered for this purpose. In this section, we consider two attributes of the local terrain that contribute to its traversability, namely local obstacles and surface softness, as described below.

2.1 Local Obstacles

Local obstacle is the generic name that refers to large rocks ("positive" obstacles) or deep ditches ("negative" obstacles) that are impassable by the robot. A wheeled mobile robot can go over obstacles that are in certain proportion to its wheel diameter. For instance, a JPL-preferred mobility mechanism called the rocker-bogie design can climb obstacles 1.5 times its wheel diameter [13]. Smaller obstacles are therefore not considered as hazards to the robot mobility. Larger obstacles, however, impede the robot motion and must be considered. The presence of large obstacles can be detected in real-time by proximity sensors (for rocks) and cameras (for ditches) mounted on the robot [10]. Different types of proximity sensors can be used for this purpose, ranging from low-resolution infra-red sensors to high-resolution laser range-finders [see, e.g., 14], and the range of operation of these local sensors is typically 0.2 meters to 1 meter. Each local sensor (such as proximity sensor or camera) measures the distance d_o between the robot and the *closest* obstacle within its range of operation, and this information is continuously updated during robot motion. The closest obstacle distance d_o is represented by three fuzzy sets with the linguistic labels $\{VERY - NEAR, NEAR, FAR\}$, with the trapezoidal membership functions shown in Figure 1a. Note that we can have different definitions of these membership functions for the front obstacle and the side (left and right) obstacles so that

front and side navigation behaviors will have different sensitivities. Observe that *precise* measurement of the obstacle distance is *not* needed, because of the multi-valued nature of the linguistic fuzzy sets used to describe it.

2.2 Surface Softness

Local surface softness directly affects the traction of a mobile robot traversing a challenging terrain. Different ground material, whether soft sand, loose gravel, or compacted soil, exhibit different contributions to the robot's ability to travel effectively on the surface. For example, extremely gravely surfaces cause excessive wheel slippage, and thus are deemed unsafe for traversal. Soft sandy surfaces may cause the robot to sink, and should also be avoided. Surface material properties thus contribute directly to robot safety and must be included in local terrain assessment.

There are several methods for assessing the surface softness in close proximity of the robot. One concept is a non-contact sensor that consists of a pneumatic probe which will output a puff of air toward the ground surface and a laser displacement sensor that will detect the associated ground displacement. For "soft" ground, the detected surface displacement will be very large and for "hard" ground, the displacement value will be minimal. Another concept is a small force sensor carried by a simple mechanism attached to the robot that makes physical contact with the nearby surfaces and senses the resulting contact forces [15] (analogous to a person feeling the surface with a walking stick). These sensors will enable the robot to distinguish hazardous soft sandy region from safe hard compacted soil. Yet another method to determine the surface type, and as a result the surface softness, is based on visual texture analysis using neural networks [10]. This is a two-step approach; in the first step a neural network classifier is trained off-line using a set of known sample texture prototypes. In the second step, the trained neural network is used to recognize the ground texture acquired during run-time. The perceived surface type is then fed into a look-up table for obtaining surface softness γ. This softness factor is characterized by three fuzzy sets with the linguistic labels $\{SOFT, MEDIUM, HARD\}$, with the trapezoidal membership functions shown in Figure 1b. Again, observe that *precise* measurement of the surface softness is *not* needed, because of the multi-valued nature of the linguistic fuzzy sets used to describe it.

2.3 Local Traversability Index

Once the characteristics of the local terrain are obtained in terms of the closest obstacle distance d_o and local surface softness γ, this information can be incorporated into a single index of local traversability τ_l. This index is represented by four fuzzy sets with the linguistic labels $\{POOR, LOW, MODERATE, HIGH\}$, with the trapezoidal membership functions shown in Figure 1c. The relationship between the Local Traversability Index τ_l and the obstacle distance d_o and surface softness γ is expressed by a set of simple linguistic fuzzy logic rules. These rules are summarized in Table 1, with d_o and γ as two inputs and τ_l as the single output.

Observe that by utilizing fuzzy logic, the outcome of the rule set τ_l is *not* dependent on *exact* measurements of the obstacle distance d_o and the surface softness γ. This feature allows *robust* assessment and classification of the local terrain using imprecise sensors. This is because in the fuzzy logic formulation, the input variables d_o and γ are allowed to vary over a range of values without altering the output variable τ_l.

3 Regional Traversability Analysis

The regional traversability covers a zone of typically up to 5 meters away from the mobile robot for a small-sized robot. The physical and geometrical qualities of the terrain segment within this zone determine its ease-of-traversal by the mobile robot. Several characteristics of the terrain can be considered for this purpose. The most notable ones are the terrain slope and roughness. These two characteristics are extracted from video image data obtained by the stereo cameras mounted on the mobile robot, as described below [10].

3.1 Terrain Roughness

The terrain roughness can be defined in several different ways. In this paper, we choose an in-tuitive approach by defining the region roughness in terms of the sizes and concentrations of rocks in that region. The rock detection algorithm [10] is applied to stereo camera images of the viewable scene to identify target objects located on the ground plane using a region-growing method [16]. Rock sizes are classified as $\{SMALL, LARGE\}$ depending on their pixel counts relative to a user-defined threshold. Rock concentrations are classified as $\{FEW, MANY\}$ depending on the number of rocks in the region relative to a user-specified limit. The terrain roughness is then determined based on the rock sizes and concentrations in the region, and is represented by the four linguistic fuzzy sets $\{SMOOTH, ROUGH, BUMPY, ROCKY\}$. Table 2 summarizes the definition of terrain roughness in terms of rock sizes and concentrations using a set of four simple linguistic fuzzy logic rules.

3.2 Terrain Slope

To obtain the terrain slope from a pair of stereo camera images, we first calculate the real-world Cartesian x, y, z components of the ground plane boundary [10]. Tsai's camera calibration model [17] is used to derive the relationship between the camera image and the real-world object position for a single camera. The images from both cameras are then matched in order to retrieve 3D information. The average slope value α is then determined using the equation $\alpha = \frac{1}{N} \sum_i^N atan2(z_i, x_i)$, where N is the number of points viewable in both images. The terrain slope α is represented by the four linguistic fuzzy sets $\{FLAT, SLANTED, SLOPED, STEEP\}$.

3.3 Regional Traversability Index

Once the characteristics of the viewable scene are extracted, the terrain traversal must be assessed. To accomplish this task, we have developed a set of fuzzy logic rules which assess the traversability of the terrain based on the characteristics present in the given image data set. The Regional Traversability Index τ_r encapsulates multiple terrain characteristics into a single index and succinctly quantifies the ease-of-traversal of the terrain by the mobile robot.

In order to characterize the terrain, the terrain characteristics are first converted into linguistic representations using fuzzy sets. These sets allow each terrain characteristic to be represented based on *grades* of membership to user-defined linguistic fuzzy sets. The membership functions of these sets are then used in a set of fuzzy logic rules to infer terrain traversability. These simple fuzzy relations are summarized in Table 3. The output from the rule base is the Regional Traversability Index τ_r which represents the relative level of safety associated with traversing the viewable area. This index is represented by four fuzzy sets with the linguistic labels $\{POOR, LOW, MODERATE, HIGH\}$. By utilizing fuzzy logic, the user can specify rules that are not dependent on *exact* measurements of the terrain characteristics, thus allowing *robust* analysis of the terrain.

4 Global Traversability Analysis

In previous sections, we present local and regional traversability analyses using on-board sensors, with ranges of resolution typically 0.5 meters and 5 meters for a small-sized robot. In this section, a different type of terrain traversability is discussed which is based on the *terrain map* and can operate in the global scale in tens of meters resolution, well beyond the robot's sensing envelope.

4.1 Global Traversability Map

The Global Traversability Map classifies the terrain segments based on how difficult and unsafe each segment is for traversal by the mobile robot. The map building process involves two steps. We first identify relevant topographic terrain features (such as ravines, mountains, and valleys) as observed in aerial imagery or obtained from land surveys. Various image-based techniques can be used to identify these relevant terrain features. For example, to identify ravines, an approach can be utilized which locates curving linear features embedded in the image using edge-detection techniques [18]. For identifying mountains and hills, the peaks and valleys can be found based on contour lines [19].

Once the relevant topographic terrain features are extracted, they are fed into a linguistic rule set for constructing the Global Traversability Map. This rule set assigns a traversability index to each terrain segment that reflects the global-scale terrain quality for traversal. The segment classification can be performed using four fuzzy sets with the linguistic labels $\{POOR, LOW, MODERATE, HIGH\}$, as in Sections 2-3. Each traversability class designates the traversal risk/difficulty associated with that segment, namely unsafe, moderately-unsafe, moderately-safe, and safe. For example, IF *Mountain Slope* is STEEP, THEN *Traversability* is POOR. In a similar manner, a large gorge can easily be designated as untraversable, and thus will receive a POOR traversability index; whereas a mountain or a hill depending on the slope may receive a POOR to MODERATE traversability index. Note that while the mountain peak has a POOR traversability value, as we move down slope to the foothills, the traversability can change to LOW and MODERATE. Therefore, the mountain can be characterized by a set of three concentric circles with different traversability indices as shown in Figure 2a.

We define a fixed map-based coordinate frame-of-reference, with the (x, y) plane on the terrain and the z-axis perpendicular to the terrain. At any time, the robot is aware of its own coordinates on the traversability map using robust position estimation and self-localization methods. The robot position needs to be established relative to the global map features [20]. To represent the Global Traversability Map to the robot navigation system, the user can, for instance, choose one of the following two representations:

- *Traversability Regions:* Each segment is approximately bounded by the coordinate inequalities $\{X_{min} \leq x \leq X_{max}, Y_{min} \leq y \leq Y_{max}\}$, as shown in Figure 2b. This, in effect, defines the rectangular area in the terrain map that is occupied by the particular feature. Alternatively, each segment is enclosed by a geometric shape such as a circle. The enclosing circle is mathematically described by $(x - a)^2 + (y - b)^2 = r^2$, where (a, b) are the center coordinates and r is the radius–an example is shown in Figure 2b.

- *Traversability Grid:* We overlay on the map an MxN grid composed of MN equal-sized grid cells, where M and N are user-defined numbers chosen based on the map resolution and the robot footprint. Each grid cell is assigned a traversability index that reflects the *minimum* index of all terrain segments occupying that cell (see Figure 2c).

Alternatively, the fuzzy map representation method [21] can be used, where the locations of the traversability regions on the map are expressed by a set of fuzzy logic statements.

The procedure for generation of the Global Traversability Map is carried out *off-line*. Once this map is generated, its mathematical model is uploaded in the memory of the computing platform mounted on the robot. From the robot navigation perspective, the Global Traversability Map is available to the robot navigation system *prior* to the robot movement.

4.2 Global Traversability Index

Once the Global Traversability Map is generated, we can compute the Global Traversability Index of the mobile robot in different terrain sectors at any time. For this purpose, we proceed as follows:

- Decompose the terrain available to the mobile robot into several circular sectors centered at the current robot position and having radius R_g. The value of R_g determines the *reaction distance* of the robot, and is the distance at which we wish the robot to react to the global surface features.

- For each circular sector, assign the *minimum* traversability index of the map segments contained within that sector. This can be obtained using geometric calculation of the intersections between the sector and the segments. The rationale for using the minimum index is to enhance robot safety, given the fact that the map information and terrain classification are often inaccurate and approximate.

The outcome of this procedure is the Global Traversability Index τ_g that corresponds to a particular terrain sector.

5 Conclusions

Multi-scale traversability indices are introduced in this paper for a field mobile robot operating on a challenging natural terrain. These indices quantify the difficulty/risk associated with the robot mobility at three scales of resolution. Terrain-based navigational behaviors based on traversability indices are critical components of any field robot navigation strategy. These behaviors provide a means for incorporating different terrain characteristics into the robot navigation logic. Current research is focused on implementation and field testing of the methodology described in this paper on a commercial mobile robot.

6 References

1. S. Singh, et al: "Recent progress in local and global traversability for planetary rovers", Proc. IEEE Intern. Conf. on Robotics and Automation, vol. 2, pp. 1194-1200, San Francisco, 2000.

2. A. Kelly and A. Stentz: "Rough terrain autonomous mobility", Journal of Autonomous Robots, vol. 5, no. 2, pp. 129-198, 1998.

3. R. Simmons, et al: "Experience with rover navigation for Lunar-like terrains", Proc. IEEE/RSJ Intern. Conf. on Intelligent Robots and Systems (IROS), pp. 441-446, Pittsburgh, 1995.

4. E. Krotkov, et al: "Field trials of a prototype Lunar rover under multi-sensor safeguarded teleoperation control", Proc. American Nuclear Society 7th Topical Meeting on Robotics and Remote Systems, vol. 1, pp. 575-582, Augusta, 1997.

5. A. Stentz and M. Hebert: "A complete navigation system for goal acquisition in unknown environments", Proc. IEEE/RSJ Intern. Conf. on Intelligent Robots and Systems (IROS), vol. 1, pp. 425-432, Pittsburgh, 1995.

6. D. Langer, J. K. Rosenblatt, and M. Hebert: "A behavior-based system for off-road navigation", IEEE Trans. on Robotics and Automation, vol. 10, no. 6, pp. 776-783, 1994.

7. H. Seraji: "Traversability Index: A new concept for planetary rovers", Proc. IEEE Intern. Conf. on Robotics and Automation, vol. 3, pp. 2006-2013, Detroit, 1999.

8. H. Seraji: "Fuzzy Traversability Index: A new concept for terrain-based navigation", Journal of Robotic Systems, vol. 17, no. 2, pp. 75-91, 2000.

9. A. Howard, H. Seraji, and E. Tunstel: "A rule-based fuzzy traversability index for mobile robot navigation", Proc. IEEE Intern. Conf. on Robotics and Automation, vol. 3, pp. 3067-3071, Seoul (Korea), 2001.

10. A. Howard and H. Seraji: "Vision-based terrain characterization and traversability assessment", Journal of Robotic Systems, vol. 18, no. 10, pp. 577-587, 2001.

11. H. Seraji, A. Howard, and E. Tunstel: "Safe navigation on hazardous terrain", Proc. IEEE Intern. Conf. on Robotics and Automation, vol. 3, pp. 3084-3091, Seoul (Korea), 2001.

12. H. Seraji, A. Howard, and E. Tunstel: "Terrain-based navigation of mobile robots: A fuzzy logic approach", Proc. Intern. Symposium on AI, Robotics, and Automation in Space (i-SAIRAS), Montreal (Canada), 2001.

13. D.B. Bickler: "The new family of JPL planetary surface vehicles", Proc. Intern. Symposium on Missions, Technologies and Design of Planetary Mobile Vehicles, pp. 301-306, Toulouse (France), 1992.

14. R. Volpe and R. Ivlev: "A survey and experimental evaluation of proximity sensors for space robotics", Proc. IEEE Intern. Conf. on Robotics and Automation, vol. 4, pp. 3466-3473, San Diego, 1994.

15. P.R. Sinha, Y. Xu, R.K. Bajcsy, and R.P. Paul: "Robotic exploration of surfaces with a compliant wrist sensor", Intern. Journal of Robotics Research, vol. 12, no. 2, pp. 107-120, 1993.

16. B. Horn: "*Robot Vision*", MIT Press, MA, 1986.

17. R. Y. Tsai: "A versatile camera calibration technique for high-accuracy 3D machine vision metrology using off-the-shelf TV cameras and lenses", IEEE Journal of Robotics and Automation, vol. 3, no. 4, pp. 323-344, 1987.

18. W.B. Thompson and T.C. Henderson: "IU at the University of Utah: Extraction of micro-terrain features", DARPA Image Understanding Workshop, pp. 819-824, 1997.

19. I.S. Kwoen: "Extracting topographic terrain features from elevation maps", CVGIP: Image Understanding, vol. 59, no. 2, pp. 171-182, 1994.

20. C.F. Olson: "Landmark selection for terrain matching", Proc. IEEE Intern. Conf. on Robotics and Automation, vol. 2, pp. 1447-1452, San Francisco, 2000.

21. E. Tunstel: "Fuzzy spatial map representation for mobile robot navigation", Proc. 10th ACM Symposium on Applied Computing, pp. 586-589, Nashville, 1995.

Copyright © IFAC Telematics Applications in Automation
and Robotics, Espoo, Finland, 2004

www.elsevier.com/locate/ifac

Reconfigurable Control of Formation Antennas in Earth Orbit

Fred Y. Hadaegh, Daniel P. Scharf, Scott R. Ploen and Vaharaz Jamnejad

Jet Propulsion Laboratory
California of Institute of Technology
Pasadena, CA 91109 USA
Hadaegh@jpl.nasa.gov

Abstract: An integrated control and electromagnetic/antenna formulation is presented for evaluating the performance of a distributed antenna system as a function of formation geometry. A distributed and self-organizing control law for the control of multiple antennas in Low Earth Orbit (LEO) is presented. The control system provides collaborative commanding and performance optimization to configure and operate the distributed formation system. A large aperture antenna is thereby realized by a collection of miniature sparse antennas in formation. A case study consisting of a simulation of four antennas in Low Earth Orbit (LEO) is presented to demonstrate the concept. *Copyright © 2004 IFAC*

Keywords: Formation flying, Leader/Follower, Self-organizing, Sparse antenna, Formation control, Distributed control.

1. Introduction

In recent years, the science community has been actively considering the use of distributed spacecraft for deep space and Earth science missions. One such application is to use a large number of small spacecraft in place of a large deployable antenna in order to achieve very large sparse apertures for Earth imaging (for example, at resolutions of ≈ 10 cm). Another application is the use of multiple telescopes flying in precision formation as an interferometer in deep space for stellar imaging and planet detection. A number of such missions have been proposed that offer unprecedented performance capabilities beyond the scope of any single large telescope [11,12]. Compared to their equivalent monolithic aperture counterparts, formation flying sparse antennas offer launch and deployment efficiency, and has the advantage of avoiding the structural complexity and pointing issues associated with large aperture, lightweight, antenna dishes in space.

This paper presents an integrated control and electromagnetic/antenna approach needed to realize, for the first time, distributed formation flying spacecraft antenna systems in Low Earth Orbit (LEO). The paper focuses on the core guidance and control (G&C) algorithms needed to perform parametric studies to access the impact of replacing a large monolithic space-based antenna by a collection of miniature spacecraft. Formation guidance and control design for both translation and attitude are presented in Section 3 and 4. Section 5 provides

analysis of a spatial array of antennas along with simulations. Section 6 presents a four-spacecraft sparse aperture example for evaluation of the distributed antenna system performance.

2. Sparse Antenna Guidance and Control Architecture

In general, the methodology for coordination and control of spacecraft in a formation is strongly correlated with the formation size and particular application. In a completely centralized architecture, a single master spacecraft commands all aspects of the other slave spacecraft. At the other end of the spectrum is a completely decentralized architecture in which spacecraft interact locally with other nearby spacecraft. In this latter case, formation behavior is said to be "emergent," and is similar to the schooling of fish or the flocking of birds. The defining characteristic of a decentralized architecture is that individual spacecraft do not require knowledge of the entire formation state for control.

Here we use the Leader/Follower (L/F) decentralized control architecture [5] for controlling relative spacecraft *positions* (attitude control is discussed subsequently). This architecture is robust and scaleable (e.g., individual spacecraft failures do not affect the overall formation stability and additional spacecraft can be easily added using only local control design). In the L/F architecture, individual small to medium formations (i.e., 5 to 10 spacecraft).

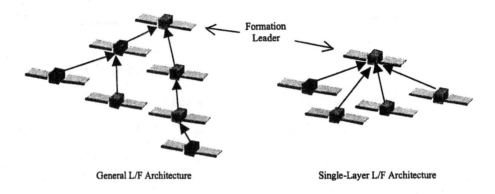

Figure 1. Two possible L/F architectures. Solid arrows indicate leader assignment.

In this case, a single-layer L/F architecture is feasible. Hence, all spacecraft follow the same spacecraft, which is the designated formation leader. For larger formations, single-layer L/F imposes restrictive inter-spacecraft communication and sensing requirements. Figure 1 shows examples of a general L/F architecture and a single-layer L/F architecture.

Absolute spacecraft attitudes are controlled independently so that individual apertures are pointed in the nadir direction. A spacecraft's nadir direction is determined using real time inertial position knowledge obtainable from GPS measurements to 10 m(1σ) accuracy (we cannot use precision centimeter level orbit determination as was used for Topex/Poseidon since this position data is delayed by up to a day) [9]. For the orbits considered, a 10 m inertial positioning error leads to a maximum nadir pointing error of 0.1 arc-minutes. Therefore, inertial positioning errors can be ignored for our purposes.

The formation guidance has a hybrid architecture; part centralized and part decentralized. The attitude guidance is decentralized; each spacecraft points its aperture in the nadir direction independently of the other spacecraft. The translational guidance is centralized. A path-planning algorithm on the formation leader plans the relative trajectories of each follower with respect to the formation leader. These trajectories are then communicated to the followers.

For a collection of spacecraft (apertures) to function cooperatively as a sparse antenna, the control system must be capable of maintaining specified relative spacecraft positions to a fraction of the antenna wavelength. As a result, scientific applications require *precision formation flying* (i.e., centimeter/arc-minute-level relative position/attitude control). Relative position requirements have been previously studied for synthetic aperture applications:

In the VHF radio frequency band (i.e., 1 to 10 m wavelengths), relative spacecraft positions must be controlled to approximately the 15 cm level. Similarly, for interferometric synthetic aperture radar applications in the L band (15 to 30 cm wavelengths), relative spacecraft positions must be controlled to approximately the 3 cm level [1]. These relative positioning requirements are consistent with current carrier differential phase GPS (CDGPS) sensors, which can measure relative positions with 2 cm (1σ) of accuracy. Attitude requirements for radar and radio frequency synthetic apertures are not as well defined [2]; a spacecraft must only point to a fraction of an aperture's beam pattern width [3], which is application dependent. In this paper, we assume that all the spacecraft are nadir-pointing (i.e., down-looking) and that the attitude control requirements are consistent with attitude sensing via CDGPS (i.e., 5 to 10 arcminute level)[6].

In summary, robust precision formation control and guidance algorithms must be developed that (1) maintain relative spacecraft positions and absolute attitudes to 5 cm and 10 arc-minutes, respectively, and that (2) reconfigure the formation using fuel-optimal, collision free trajectories. Further, these algorithms must perform over orbits with altitudes ranging from 250 to 1000 km and non-zero eccentricity.

3. Formation Guidance Design

The formation guidance algorithm has two functions: (1) planning relative positions of the follower spacecraft so that the desired electromagnetic beam pattern is attained, and (2) planning fuel-optimal, collision-free reconfiguration trajectories to form new beam patterns or balance fuel consumption. The first guidance function requires optimal aperture positioning (a genetic-algorithm based approach is presently under study), and a prescribed set of relative spacecraft positions is used for this purpose.

The second guidance function has been designed and implemented using two different algorithms. The first algorithm is applicable to formations in circular orbits, and is based on linearized Lambert targeting (LLT) using the Hill-Clohessy-Wiltshire (HCW) equations discussed below. The collision avoidance algorithm for LLT guidance is heuristic-based, and is not guaranteed to converge to collision-free trajectories nor is it optimal. However, the LLT algorithm is a quick and efficient method for calculating reconfigurations. The second reconfiguration guidance algorithm is an implementation of the linear programming (LP) algorithm of [9]. The LP algorithm is applicable to formations in eccentric orbits. However, it is optimal only when the fleet leader is fixed on a reference orbit. The LP algorithm first discretizes the control input and then minimizes the absolute value of the acceleration for a spacecraft reconfiguration. For our purposes, the main benefit of the LP algorithm is the ability to enforce state constraints for collision avoidance.

4. Formation Control Design
4.1 Translational Control
Since our primary goal is to develop a general formation controller to support sparse aperture beam pattern analysis/optimization over a wide range of formation orbits, a classical design method was chosen for developing the individual spacecraft translational control-laws. Classical design methods have straightforward robustness criteria and have proven to perform adequately even when design assumptions are violated. The control design-model is based on the HCW equations, which describe the relative (linearized) translational dynamics between a leader and follower spacecraft when they are near a circular orbit. The reference frame and variables used in the HCW equations are shown in Figure 2. The HCW frame has an origin O traveling on a circular reference orbit and coordinate axes $\hat{x}_h, \hat{y}_h$, and $\hat{z}_h$ where $\hat{y}_h$ is parallel to the circular orbit velocity, $\hat{z}_h$ is perpendicular to the orbital plane, and $\hat{x}_h$ completes the right-handed triad. The HCW frame is also rotating with constant angular velocity

$\vec{\omega}_0 = \omega_0 \hat{z}_h$. The position of the leader in the HCW frame is given by $\vec{\rho}_j$ and the position of the follower by $\vec{\rho}_i$. The position of the leader with respect to the follower, resolved in the HCW frame, is given by $\rho_{ij} = [x \quad y \quad z]^T$. When both $|\vec{\rho}_j|$ and $|\vec{\rho}_i|$ are small compared to the orbital radius, the equations of motion are

$$\ddot{x} - 3\omega_0^2 x - 2\omega_0 \dot{y} = a_x \tag{1}$$

$$\ddot{y} + 2\omega_o \dot{x} = a_y \tag{2}$$

$$\ddot{z} + \omega_o^2 z = a_z \tag{3}$$

where a_x, a_y and a_z are inertial accelerations due to all control forces and disturbances resolved in the HCW frame. constitutes a multiple-input, multiple-output (MIMO) system, and so classical MIMO design methods must be used.

In the HCW equations, the x and y motion is coupled, and the z motion is decoupled. As a result, the z motion controller can be designed using standard single-input, single output (SISO) classical design methods. However, the combined x and y motion There are a variety of classical MIMO design techniques, but generally they first attempt to make a MIMO system look approximately SISO, and then apply SISO design methods [7,8]. We adopt the sequential loop closure MIMO technique, illustrated in Figure 3. This technique generates a diagonal MIMO controller in the frequency domain of the form

$$K(s) = \begin{bmatrix} K_1(s) & 0 \\ 0 & K_2(s) \end{bmatrix} \tag{4}$$

by only considering unidirectional (i.e., hierarchical) MIMO coupling during the initial control design. After an initial controller has been obtained, the control design is iterated with full bi-directional coupling to guarantee stability.

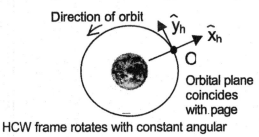

Direction of orbit $\hat{y}_h$ $\hat{x}_h$

O

Orbital plane coincides with page

HCW frame rotates with constant angular

rate ω_o

Leader, spacecraft j

$\hat{y}$ $\hat{z}$ ρ_{ij}

ρ_i

ρ_j Follower, spacecraft i

$\hat{x}$

O

Figure 2. HCW Frame and Variables used in the HCW Equations

A representative single loop control design (e.g., the z motion control loop or the $K_1(s)G_{11}(s)$ loop in Figure 3) is shown in Figures 4 and 5. Figure 4 is the Bode diagram for the representative loop transfer function. The ω_0 resonance at 2×10^{-4} Hz which is proportional to the reciprocal of the reference orbit period is apparent. Note that since the control is implemented at 1 Hz, the Bode plot only extends to 0.5 Hz.

Figure 5 shows the loop frequency response (solid line) in the L-plane (i.e., magnitude in dB versus phase in degrees). The frequency response must remain outside the dark dashed box for the standard stability margins of 6 dB and 30 degrees. The light dashed lines indicate the conditional stability boundaries (i.e., if the frequency response remains below and to the right of the green lines, then control saturation will not necessarily destabilize the controller). To obtain increased control performance, the translational controller was designed to be conditionally stable.

4.2 Attitude Control

The attitude control design model is based on Euler's equations (i.e. the balance of angular momentum) linearized about the Local-Vertical-Local-Horizontal (LVLH) frame. The LVLH frame travels on a circular orbit and rotates with the circular orbit just as the HCW frame does. In particular, the LVLH frame rotates with constant angular velocity $\bar{\omega}_{orbit} = -\omega_0 \hat{y}_L$, where ω_0 is the same as in the HCW equations. However, the LVLH frame is rotated with respect to

the HCW frame so that the $\hat{z}_L$ coordinate axis points in the nadir direction. See Figure 6.

The angular deviations in each axis are given by ε_{xL}, ε_{yL} and ε_{zL}, respectively. An "L" subscript has been included to indicate the LVLH frame as opposed to the HCW frame. The resulting attitude dynamics are then

$$I_x\ddot{\varepsilon}_{xL} + (I_y - I_z)\omega_0^2\varepsilon_{xL} + (I_y - I_x - I_z)\omega_0\dot{\varepsilon}_{zL} = \tau_x \quad (5)$$

$$I_z\ddot{\varepsilon}_{zL} + (I_y - I_x)\omega_0^2\varepsilon_{zL} + (I_x + I_z - I_y)\omega_0\dot{\varepsilon}_{xL} = \tau_z \quad (6)$$

$$I_y\ddot{\varepsilon}_{yL} = \tau_y \quad (7)$$

where I_x, I_y and I_z are the principal moments of inertia about the spacecraft center of mass, and τ_x, τ_y, and τ_z denote the control and disturbance torques resolved in the LVLH frame. Two implicit assumptions in Equations (5)-(7) are that the principal axes of the spacecraft coincide with the body frame and that the body frame coincides with the LVLH frame when the attitude control error is zero.

Note that Equations (5) and (6) are coupled, whereas (7) is decoupled. This structure is identical to the structure of the translational control design model given in Equations (1)-(3). Although, in this case the $\hat{x}_L$ and $\hat{z}_L$ motions are coupled. Since the design process is essentially the same, we only report the results of the attitude control design.

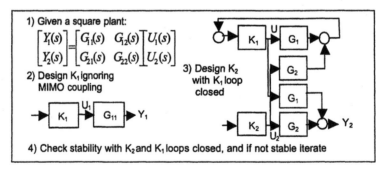

Figure 3. Sequential Loop Closure Design Technique

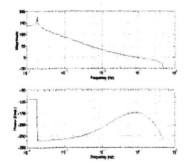

Figure 4. Bode Diagram of Representative

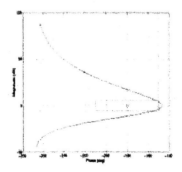

Figure 5. L-Plane Diagram of Representative Controller Loop Transfer Function

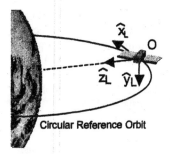

Figure 6. Local-Vertical-Local-Horizontal Frame

5. MATLAB Implementation

In the first implementation we employed the full cosine pattern for element pattern representation. Both analytical closed-form solutions as well as a numerical integration method were utilized. The input to the program includes the position of each element in space and the orientation of the array. In its present form, the program calculates the peak gain of the array, and plots pattern cuts corresponding to $\phi=0$ deg. (x-z plane), $\phi=90$ deg. (y-z plane), or any other specified azimuth angle ϕ. In addition, three-dimensional plots and contour plots of the beam pattern in u-v space, where $u=sin(\theta)cos(\phi)$, $v=u=sin(\theta)sin(\phi)$, and θ denotes the co-latitude are provided. In the final implementation of the program, a general three-dimensional array of elements with arbitrary locations and orientations were considered. Full cosine, as well as half-cosine and lambda element pattern representations, were included. The input to the program includes the location of each element in space as well as the Eulerian angles describing its orientation. The program calculates the peak gain of array, and plots pattern cuts corresponding to any specified azimuth angle ϕ. In addition three-dimensional plots as well as contour plots of the beam pattern in u-v space are provided. As an example, the beam contour plots in u-v space of a two-element array ($x=\pm1\lambda$, where λ is the wavelength) with variations of element positions in x, y, and z directions are shown in Figure 6.

6. G&C Simulation Architecture

An integrated guidance and control simulation testbed was developed that allows direct visualization of the coupling between orbital motion and the three-dimensional (spatial) antenna array pattern. The testbed was used to perform computer simulations to study the dynamic behavior of the distributed antenna formations, and assess the performance of the separated spacecraft antenna system. The G&C algorithms described above have been integrated with

the electromagnetic/antenna field computations to analyze a formation of N distributed antennas in LEO. The geometric and mass properties of the spacecraft can be varied. The simulation architecture has blocks for formation estimation, guidance and control. Different formation scenarios are addressed by modifying the formation guidance blocks in each of the spacecraft. An inertial vector propagator is used to update the nadir direction and there are also blocks for mode commanding and thrust allocation. Currently, inter-spacecraft communication is assumed ideal (i.e., no delays). See Figures 7 and 8.

7. Conclusion

This paper introduced modeling, and G&C methodologies for a set of antennas flying in formation in LEO. The electromagnetic/antenna performance of a distributed and self-organizing formation was evaluated through a control system that achieved collaborative commanding and performance optimization to configure and operate the distributed formation system. A case of four antennas in Low Earth Orbit (LEO) was presented and it was shown that proper configuration and orbital positioning of antennas could lead to unprecedented antenna system performance capability.

The simultaneous control and coordination of individual spacecraft is a very complicated task. Specifically, maintaining precision position and attitude tolerances while coordinating RF excitations for various elements in a spatially separated environment is a non-trivial task. These areas must be studied in greater detail to provide feasible implementation techniques.

Acknowledgments
The work described in this paper was carried out at the Jet Propulsion Laboratory, California Institute of Technology, under contract with the National Aeronautics and Space Administration.

References
[1] Goodman, N.A., Lin, S.C., Rajakrishna, D., and Stiles, J.M., "Processing of Multiple-Receiver Spacebourne Arrays for Wide-Area SAR," *IEEE Trans. Geoscience and Remote Sensing*, Vol. 40(4), pp. 841-852, 2002.
[2] Corazzini, T., Robertson, A., Adams, J.C., Hassibi, A., and How, J.P., "Experimental Demonstration of GPS as a Relative Sensor for Formation Flying Spacecraft," *J. of the Institute of Navigation*, Vol. 45(3), pp. 195-207, 1998.
[3] Moccia, A., and Vetrella, S., "A Tethered Interferometric Synthetic Aperture Radar (SAR) for a

Topographic Mission," *IEEE Trans. Geoscience and Remote Sensing*, Vol. 30(1), pp. 103-109, 1992.

[4] Massonnet, D., "Capabilities and Limitations of the Interferometric Cartwheel," *IEEE Trans. Geoscience and Remote Sensing*, Vol. 39(3), pp. 506-520, 2001.

[5] Wang, P.K.C., and Hadaegh, F.Y., 'Coordination and Control of Multiple Microspacecraft Moving in Formation," *J. of the Astronautical Sci.*, Vol. 44(3), pp. 315-355, 1996.

[6] Crassidis, J.L., Lightsey, E.G., and Markley, F.L., "Efficient and Optimal Attitude Determination Using Recursive Global Positioning System Signal Operation," *J. Guidance, Control, and Dynamics*, Vol. 22 (2), pp. 193-201, 199.

[7] Maciejowski, J.M., <u>Multivariable Feedback Design</u>, Addison-Wesley Publishing Company: Wokingham, England, 1989.

[8] Hovd, M., Braatz, R.D., and Skogestad, S., "SVD Controllers for H_2-, H_8- and μ-Optimal Control," *American Control Conf.*, 1994.

[9] Tillerson, M., Inalhan, G., and How, J.P., "Coordination and Control of Distributed Spacecraft Systems Using Convex Optimization Techniques," *Int. J. of Robust and Nonlinear Control*, Vol. 12, pp. 207-242, 2002.

[10] Singh, G. and Hadaegh, F.Y., "Collision Avoidance Guidance for Formation-Flying Applications," *AIAA Guidance, Navigation, and Control Conf.*, 2001.

[11] Gendreau, K.C., White, N., Owens, S., Cash, W., Shipley, A., and Joy, M., "The MAXIM X-ray Interferometry Mission Concept Study," *36th Liege Int. Astrophysics Colloquium*, Liege, Belgium, 2001.

[12] Beichman, C.A., "The Terrestrial Planet Finder: The Search for Life-Bearing Planets Around Other Stars," *SPIE Conference on Astronomical Interferometry*, Kona, Hawaii, 1998.

[13] Bauer, F.H., Hartman, K., and Lightsey, E.G., "Spacebourne GPS Current Status and Future Visions," *IEEE Aerospace Conf.*, 1998.

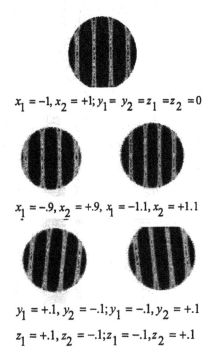

$$x_1 = -1, x_2 = +1; y_1 = y_2 = z_1 = z_2 = 0$$

$$x_1 = -.9, x_2 = +.9, x_1 = -1.1, x_2 = +1.1$$

$$y_1 = +.1, y_2 = -.1; y_1 = -.1, y_2 = +.1$$

$$z_1 = +.1, z_2 = -.1; z_1 = -.1, z_2 = +.1$$

Figure 7 Beam Contour Variations

Figure 7. Simulation testbed GUI

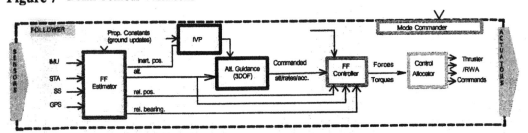

Figure 8. Formation G&C architecture

AUTHOR INDEX

Title/Year of publication	Editor(s)	ISBN
2002 continued		
Periodic Control Systems (W)	Bittanti & Colaneri	0 08 043682 X
Modeling and Control in Environmental Issues (W)	Sano, Nishioka & Tamura	0 08 043909 8
Computer Applications in Biotechnology (C)	Dochain & Perrier	0 08 043681 1
Time Delay Systems (W)	Gu, Abdallah & Niculescu	0 08 044004 5
Control Applications in Post-Harvest and Processing Technology (W)	Seo & Oshita	0 08 043557 2
Intelligent Assembly and Disassembly (W)	Kopacek, Pereira & Noe	0 08 043908 X
Adaptation and Learning in Control and Signal Processing (W)	Bittanti	0 08 043683 8
New Technologies for Computer Control (C)	Verbruggen, Chan & Vingerhoeds	0 08 043700 1
Internet Based Control Education (W)	Dormido & Morilla	0 08 043984 5
Intelligent Autonomous Vehicles (S)	Asama & Inoue	0 08 043899 7
2003		
Proceedings of the 15th IFAC World Congress 2002 (CD + 21 vols)	Camacho, Basanez & de la Puente	008 044184 X
Modeling and Control of Economic Systems (S)	Neck	0 08 043858 X
Mechatronic Systems (C)	Tomizuka	0 08 044197 1
Programmable Devices and Systems (W)	Srovnal & Vlcek	0 08 044130 0
Real Time Programming (W)	Colnaric, Adamski & Wegrzyn	0 08 044203 X
Lagrangian and Hamiltonian Methods in Nonlinear Control (W)	Astolfi, Gordillo & van der Schaft	0 08 044278 1
Intelligent Control Systems and Signal Processing (C)	Ruano, Ruano & Fleming	0 08 044088 6
Guidance and Control of Underwater Vehicles (W)	Roberts, Sutton & Allen	0 08 044202 1
Analysis and Design of Hybrid Systems (C)	Engell, Gueguen & Zaytoon	0 08 044094 0
Intelligent Manufacturing Systems (W)	Kadar, Monostori & Morel	0 08 044289 7
Control Applications of Optimization (W)	Gyurkovics & Bars	0 08 044074 6
Fieldbus Systems and Their Applications (C)	Dietrich, Neumann & Thomesse	0 08 044247 1
Intelligent Components and Instruments for Control Applications (S)	Almeida	0 08 044010 X
Modelling and Control in Biomedical Systems (S)	Feng & Carson	0 08 044159 9
2004		
Advances in Control Education (S)	Lindfors	0 08 043559 9
Robust Control Design (S)	Bittanti & Colaneri	0 08 044012 6
Fault Detection, Supervision and Safety of Technical Processes (S)	Staroswiecki & Wu	0 08 044011 8
Technology and International Stability (W)	Kopacek & Stapleton	0 08 044290 0
System Identification (SYSID 2003) (S)	Van den Hof, Wahlberg & Weiland	0 08 043709 5
Control Systems Design (C)	Kozak & Huba	0 08 044175 0
Robot Control (S)	Duleba & Sasiadek	0 08 044009 6
Time Delay Systems (W)	Garcia	0 08 044238 2
Control in Transportation Systems (S)	Tsugawa & Aoki	0 08 0440592
Manoeuvring and Control of Marine Craft (C)	Batlle & Blanke	0 08 044033 9
Power Plants and Power Systems Control (S)	Lee & Shin	0 08 044210 2
Automated Systems Based on Human Skill and Knowledge (S)	Stahre & Martensson	0 08 044291 9
Automatic Systems for Building the Infrastructure in Developing Countries (Knowledge and Technology Transfer) (W)	Dimirovski & Istefanopulos	0 08 044204 8
Intelligent Assembly and Disassembly (W)	Borangiu & Kopacek	0 08 044065 7
New Technologies for Automation of the Metallurgical Industry (W)	Wei Wang	0 08 044170 X
Advanced Control of Chemical Processes (S)	Allgöwer & Gao	008 044144 0

Customers wishing to obtain details of all available IFAC volumes, should contact their nearest Elsevier office or check the IFAC Publications website (www.elsevier.com/locate/ifac).